Payoff Artillery - WWII

The wide-ranging combat of the battalion that fought under five armies and eight divisions in the European Theatre

Copyright 1993 by Bull Run of Vermont, Inc.
with research forest on Bull Run of central Vermont

Manufactured in the United States of America
First printing

This book was printed by Queen City Printers Inc., Burlington, Vermont

Payoff Artillery - WWII

The wide-ranging combat of the battalion
that fought under five armies and eight
divisions in the European Theatre
by Frank H. Armstrong, Ph.D.

Library of Congress Catalog Card Number preassigned 93-071275
ISBN 0-9632448-1-7

Other books by the same author:
Window Seat—An aerial perspective of America's Forests
ISBN 0-9632448-0-9
Published by Bull Run of Vermont, Inc. 1992.
$11. + $3. P&H

Table of Contents

PREFACE
The deeper meaning of the title of this book

Allied artillery accounted for the overwhelming majority of German casualties on the western front. Yet, artillery troops only constituted 9.4 percent of U.S. troop strength. One battalion of Theatre Artillery attained an astounding record of performance. This "payoff" resulted from the fortuitous assemblage of 487 men and officers who were by and large serious, dedicated, highly motivated, and unified in their sense of purpose. The word "Retribution" gradually was woven into their purpose as the battalion observed the atrocious behavior by one nation that was well on the road to world domination.

Terminology

The terms "company", "battery", and "troop" indicate the basic Army unit, usually commanded by a captain. Our cavalry branch, (mechanized or horse) has used the term "troop" whereas artillery has used the term "battery". In some cases the terms can be interchanged. Other branches, such as infantry, use the term "company".

Battalions are the next higher level. The term "battalion" is used by all branches except mechanized cavalry which uses the term "squadron".

Artillery regiments and brigades were converted to the more flexible battalions and artillery groups before 1944 excepting for one brigade.

The customary designation of army corps units is with Roman numerals. Armies, which are superior to corps, are designated with numbers.

The Meuse River in France and Belgium is called the Maas River in Holland. They are one and the same.

In World War II, "light artillery" referred to those units equipped with either 105mm howitzers or 75mm pack howitzers. "Medium artillery" connotated units equipped with 155 mm howitzers, or 4.5-inch guns. "Heavy artillery" included the 155mm gun, eight-inch howitzer, eight-inch gun, and 240mm howiters. In the case of the 155mm and the 8-inch weapons, the howitzers had shorter tubes, shorter ranges, and also were more accurate. The guns had longer tubes, longer ranges, and greater dispersion. The eight-inch howitzer, having the smallest dispersion, was our most accurate artillery piece.

The 1943 experiment in training sequencing

The fiftieth anniversary of the wartime activity of one of the most unique battalions involved in World War II is an appropriate time to consider their achievements. The non-divisional battalion was originally designated as experimental, because their filler personnel arrived directly from the reception center. It was an experiment in training sequencing. The null hypothesis was, there is no difference in combat performance between non-divisional units where the fillers received individual (boot-camp) training at replacement training centers and units where fillers were sent direct to their companies/batteries from the reception center. In the latter case, the fillers received all of their training from within the company/battery. If the null hypothesis were rejected, indicating there was a difference, then the researcher would attempt to prove which sequencing resulted in better performance. Unfortunately, the experiment was never concluded. Despite the astounding success of this one battalion no conclusions can be reached solely on the basis of one unit.

The official 1945 battalion history

At the conclusion of the war, the 283rd Field Artillery Battalion Commander appointed a committee to publish a unit history. Unfortunately, the chairman was felled by Polio. Nevertheless, the committee continued and published a unit history which includes unit photographs at the end of the war, and enlisted men's names and addresses. Unfortunately, officers' addresses were not included. The publication, printed by an unknown French firm, is not copyrighted and does not have numerical identification. One copy is on file with the U.S. Army Military History Institute at Carlisle Barracks, Carlisle, PA 17013-5008. Bull Run of Vermont, Inc., publisher of this book, can provide reproductions at a cost of $20.

Background knowledge for the tactics used

The author has continuously expanded his knowledge of the background for the experiences of the artillery battalion in which he served as a forward observer. For example, combat troops were not always informed of the reasoning for certain decisions. We experienced two delays of several days duration. The reasons are now clear. In fact, higher headquarters seemed reluctant to impart unneccesary information down to troop level because of the possibility that troops might be captured and thence reveal classified information. Members of the battalion, who reviewed drafts of this book, commented that all they knew at the time was we moved a great deal and, now finally, they had the background.

(NOTE: In the Korean War, American soldiers received far more information on the tactical situation. Frequently there were evening meetings of key non-commissioned officers where the captain revealed highly classified information such as the plan for the invasion at Inchon two days prior to the actual invasion.)

The case of the 1944 incident at Millen Castle, in Holland, has taken years to research. German western-front Wehrmacht unit histories are scarce. None have been located concerning October 1944 operations in the Limburg Province of the Netherlands. General von Zangen, who commanded the German 15th Army, wrote an after-action report after the war. Most of his report concerns actions starting shortly after the Millen affair. There are many German unit histories concerning the Russian front. Complications in recalling various operations emanated from war-time American security rules forbidding diaries and cameras. However, a few members of the battalion overlooked these rules and, as will be evident in the ensuing pages, we are indebted to them today.

The author re-visited Millen, Holland in 1957, and photographed the site. He also discussed events with local citizenry. Life, and especially historical research, was severely disrupted in Holland during the war. Holland was the only country which was under the authority of the German SS. The people suffered incredibly. When Germany surrendered, some of Holland was still occcupied by German forces. It took many years, but by 1993, Dutch historians had amassed considerable information on the war. They have been most cooperative.

Credits for contibutory information

The author has established contact with at least ten former officers, 111 men, and 88 kinfolk of our KIA and deceased. A great deal of this amazing communication, after fifty years, can be credited to Sergeant James M. Parker and his wife, Miriam. The author is indebted to the 39 of them who elected to review various drafts of this book and for their contribution to the content of the book. The Battalion Executive Officer, Major Fred Dieterle, the Operations Officer, Major (at the time) John D. Sapp, Lieutenant William Grohne, William Davis, Robert Pensock Corporal George Hoy, Sergeant George Mertz of an earlier battalion, and the former Dutch resistance youth, Jacques Briels, have been most helpful. Jacques now lives in Ogden, Utah. Dinah Lee, of Tempe, Arizona, a daughter of Lieutenant Fred Whitt (deceased) has been most helpful in providing his diary and photographs. The Commander of Troop "A" of the 125th Cavalry has also reviewed a draft. Seldom has a wartime unit history been reviewed and written by so many of the participants. Their names will be obvious in the story.

The author is indebted to the many editors of Pennsylvania's 79 daily newspapers who assisted in locating individual veterans and kinfolk. There also were three interested Pennsylvanians who undertook the challenge of locating certain kinfolk and veterans. They had had no prior contact with our battalion, but were motivated by good citizenry.

The author is also indebted to the following for their assistance in locating people, their helpful comments, and their information.

David L. Benton, III. Brigadier General U.S. Army, Asst. Commandant U.S. Army Artillery School.

H Th G van den Berk, Holtum, Netherlands.

Dr. Boyd L. Dastrup, Command Historian, U.S. Army Field Artillery Center and Fort Sill.

Brigadier General Robert F. Foley, Commandant, United States Military Academy.

Editor, Limburgs Dagblad, Heerlen, Netherlands.

David E. Johnson, Lieutenant Colonel U.S. Army, Deputy Chief Historian The Center of Military History, Washington D.C.

M. Meuwissen, Heemkunde Vereniging Nieuwstadt, Netherlands.

Dan A. Nettling. Major U.S. Army. Coordinator, World War II Commemoration, U. S. Army Military History Institute.

Nix, Jack P. Jr., Brigadier General, Assistant Commandant United States Army Infantry School who rescued the story from oblivion.

William A. Stofft, Major General, U.S. Army, Commandant United States Army War College.

Fred B. Walters, Press Secretary, Department of Military Affairs, Commonwealth of Pennsylvania

C.E.M. Strik-Zijlstra, Sectie Militaire Geschiedenis, Den Haag, Netherlands.

The cover photograph of Millen Castle is through the courtesy of BV UITGEVERSMAATSCHAPPIJ TIRION, Lt. Generaal van Heutszlaan in Baarn, Netherlands. The photos are from page 265 and 266 of "Kijk op Limburg".

Heemkunde Vereniging Nieuwstadt, and their photographer, W.C. Aalbregt made new photographs of Millen Castle for me which are greatly appreciated.

Captain Jaydee I Dodson, the real hero of Chapter 3, died some years ago but his widow has maintained contact with both the Armstrongs and the Grohnes. Colonel Hugh McDonald died about ten years ago without knowing about the publication of his superb leadership. We are all indebted to him for leadership in the American style.

I am also indebted to the authors who are listed, along with their writings, or other communications, in the Bibliography.

INTRODUCTION
Between Wars

World War I ended on 11 November 1918. The following 21 years included a few years of relative prosperity, followed by the most severe economic depression the world has ever known. The depression was climaxed by a devastating world war involving the entire globe, from Anchorage to Capetown and from Singapore to Murmansk. More than half of the World's population were residents of one of the 56 nations which were officially at war (Willmott 1989:p.474).

Childhood Adversity of the Great Depression

American children born in the early 1920's were six to ten years old when they began noticing home evictions, soup lines, bank-run lines, oatmeal for breakfast and dinner, egg-salad sandwiches for school lunches, and separate counters at Woolworths for home shoe-repair kits. There was no governmental safety net. Social organizations provided some assistance but this was minimal and did not include used-clothing banks. By and large the only security was that which could be found within expanded families.

Hot school lunches were available for those kids with money, essentially meaning for those whose dad had a job. New entrants into public schools included children who had been attending private schools. There was no adjustment problem because they soon found others from "The Academy". Clothes were worn 'till they were ready to be sold to the "rag man" when he made his monthly rounds. Used newspapers were sold to the paperman. Scraps of iron were sold to the scrap-iron man. The Japanese were paying good prices for scrap iron. Apparently, they were building all sorts of boats. Re-cycling was in full bloom. It was not legislated, but was economically motivated.

South Philadelphia's children were among the poorest. Their parents had to develop their own opportunities. The adversity which these kids underwent is not well documented because pride precluded discussion of family problems. Families were large, but only the strong survived. In response to a query from a mother who had four boys who fell ill to most every imaginable disease, a leading Philadelphia pediatrician (Creighton Turner), said "All four of yours have survived, whereas most of your neighbors have lost some of their offspring."

Throughout the world children born in the early 1920's were undergoing adversity which, as it turned out, hardened them in preparation for a highly challenging early career. In 1930 our public school experienced overcrowding and there was a redistricting which created the need for me to walk two and a half miles each way to school in all kinds of weather. At the time, I did not fully appreciate the preparation that this adversity was doing for my future career. Neither did I realize that rural Pennsylvanian kids were walking a greater distance.

> I was raised on a farm. My Dad was a rural mail courier. I walked four miles to high school. Army training was easy for me (Dotterer 1994).

The Nazis Cope with the Economic Depression

German unemployment had been 26 percent of the labor force in 1933 when the Nazi party took over the government. Creation of jobs became their immediate objective. Measures adopted included tax reductions, investment in public works, and money creation (Mills, ed. 1993). The growing governmental financial deficit was little concern because they anticipated winning the forthcoming war. The German economic stagnation came to an end. Gross domestic product rose 6.3 percent in real terms in the first year under the Nazis. Inflation was checked by price and wage controls. The goal of rearmament was deferred for a few years, but by 1938 armament expenditures accounted for 74 percent of public investment (Mills ed. 1993). Zimmerman had a family store at the corner of Chew Street and Pleasant Street in Germantown, Philadelphia. He was confronted with bankruptcy. One dark night in late 1933, the family absconded to Germany where there were jobs. Doubtlessly, other families of German descent also placed employment above freedom.

Family life suffers from booze

There were the prohibition years until 1933. The one "speak easy" in our neighborhood offered candy to children on Halloween although many parents forbade their children from going to that "awful place". It seemed to me that prohibition had a beneficial influence on our human resources because although beer and whiskey could be obtained from one of the bootleggers, it was costly and not available by the glass. Repeal of prohibition turned bootleggers and speakeasy operators into honest men overnight. Most everyone of them established a tap room (bar) or night club within a week. Beer could be had for five cents a glass, further enhanced by gratuitous hard boiled eggs, cheese, and crackers. The initial low prices for booze, at the newly opened taprooms, were intended to build a large and loyal clientele. They succeeded. Drunks flourished on street corners, and family-life suffered. I did have harsh words for Les, the bartender, who helped my father on the way towards alcoholism.

During the war years, Dad became more temperate, although there was a daily walk to the taproom to be sure his cronies were aware of his son's doings. Some kids developed an aversion to booze in response to alcoholism within the home.

The U.S. Army of the 1930's

The 1930's U.S. Army was not equipped for either a defensive nor an offensive war. It was one of the least prepared armies of all the industrialized nations. The author enlisted in June 1939 and, in this book, he cites actual cases where the lack of military research and development caused American deaths.

In the ensuing war, there were many instances where German troops had better equipment than we did. If German development of jet aircraft, and/or V-I and V-II rockets, had been a few months ahead of the actual schedule the Normandy invasion would have been far more costly in terms of our most valuable resource — America's Finest.

Revisionists

Revisionists have been targeting World War II for more than ten years. Their program has recently intensified. Some of them would have you believe the Holocaust never occured. Others would have you believe American forces maltreated disarmed enemy forces in the large detention centers at the close of the war. Some suggest German soldiers were inherently better fighters than soldiers of other nationalities. And some revisionists would have you believe it was really Hitler's war, and German people did not wholeheartedly support World War II. If you have considered any of the revisionist's theories it is time you re-examined what really happened as seen by the author. He was there.

Chapter 1
OBSOLETE-EQUIPMENT
HAMPERED TRAINING
Darkening war clouds

I was introduced to the war scene in 1931, at the age of ten. One of my father's acquaintances, Major General Smedley Butler (USMC), became involved in a case which concerned relationships between the Italian, Benito Mussolini, and our State Department. On 19 January 1931, Butler was honored at a private luncheon at the Contemporary Club of Philadelpha. He told of an incident which reflected adversely on Mussolini. The U. S. Secretary of State (Cordell Hull) wired Mussolini an apology and Butler was afforded the opportunity of either an apology to Mussolini or a court martial. This set the tone for the next ten years when the State Department considered Mussolini to be a friend of the United States. A group of Dad's cronies met at our home and discussed the affair over mugs of home brew. I listened intently. Whereas, President Roosevelt had never met the Italian dictator he fervently hoped that Mussolini could prevail on Hitler not to take the path to war (Schmitz, David F. 1988 and Venzon 1992).

Thereafter, I avidly followed the darkening war clouds. My anger was heightened in 1933 by the Nazi assassination of Chancellor Engelbert Dollfuss of Austria. Dollfus struck me as being a decent farmer with some forestry background. The assassins were immediately pardoned for the deed. Many Austrian Nazi leaders are deeply etched in my memory. Arthur Seyss-Inquart, with thick Germanic eyeglasses and always garbed in a long black coat, typified the forces of evil. Captain Theo Habicht was one of the early Austrian Nazi leaders who willingly sold his own soul to the devil for personal advancement. Sadistic bull-necked Ernst Kaltenbrunner simply wanted complete power over other people. The German Envoy to Austria, Franz von Papen, was most anxious to please Adolffus Schicklegrubber, who had assumed the name of Hitler. Von Papen wanted to maintain his own diplomatic status at any cost. German people aspired to having large landholdings. They planned to go to war to attain "Lebensraum", or living space. German military officers were delighted at the prospects of a large army which would mean promotions.

On the other hand, there was Kurt von Schuschnigg who succeeded the assassinated Dollfus as Chancellor of Austria. Schuschnigg, a career diplomat, was criticized throughout the world for readily capitulating to Hitler and "Anschluss". The latter term was the Austrian Nazi party's invitation to Germany to occupy Austria. Schuschnigg was compelled to flee to the United States when the German Army crossed into Austria. Schuschnigg spoke at West

Virginia University after the war. A number of students were openly hostile to him because of his capitulation. Others believed Schuschnigg had no other option because Austrian Nazis were in power, and they used ruthless strong-arm tactics to attain union with Germany. Schuschnigg, the only person to have openly smoked cigarettes in Hitler's presence, blamed the Austrian Nazi party for the Anschluss. Germans were overwhelmingly invited to occupy Austria, as attested to by photographs of the event. If Schuschnigg had ordered the small Austrian army to resist the occupation, many soldiers would not have obeyed and there would have been bloodshed to no avail.

Incidentally, the Nazi political party started in Austria. The Godfather of the Nazi party was Georg Schoerner, an Austrian with political activity dating back before Hitler's birth. He instigated persecution of Jews by adding a twelfth point to his 1882 Linz Program. However, the prime objective of the German Nazis, the Kaiser, and Bismark was "Lebensraum" meaning new colonies. The term "Third Reich" meant the third attempt at unifying the German people. The first was by Bismark and the second by the Kaiser. In the third attempt, the Nazis garnered public support by persecution of Jews. Politicos have found it easier to get the support of the populace when there is a so-called enemy.

I was intensely bothered by the League of Nations allowing the atrocities which were being duly reported in our newspapers. Crystal Nacht, the Night of the Long Knives, and especially the rounding up of children for imprisonment in concentration camps, were duly reported in our news media. Some Americans claim they never realized what was going on in Germany. They have short memories. As early as June 1933 American Jews boycotted Made-in-Germany merchandise until the Hitler regime changed its tactics (Forbes, June 1, 1933). Several books have been published documenting revelations of concentration camp horrors well before the war started (Lipstadt 1986).

Adolph Hitler's carved wooden horse

Gottlieb Kottmann, an expert woodcarver and a member of the Nazi Brown-Shirted SA, was one of the many Germans enamored by Hitler. In 1933 he presented Hitler with a wood carving of a mare horse with her foal. The mare is a heavily built Belgian work horse. She has a shock of hair over her forehead which uncannily resembles Adolph Hitler. Reputedly, the mare was Hitler and the foal was Germany. On the underside he carved

Dem Fuher
Adolf Hitler
Zugeeignet Marz 1933
Gottlieb Kottman — SA

Hitler was delighted with this forerunner of Nazi art. It remained on his mantel piece at the Berchtesgaden retreat until 5 May 1945. A German postage stamp featured the wood carving. The carving had come to represent the relationship between Hitler and the German people.

Hitler-Youth invasion of Holland repelled

In early 1939, Jacques Briels went ice skating on the moat which surrounded the newer Millen Castle, which hereafter I will refer to as the manor house in order to differentiate it from the original castle. The original castle dates back, at least, to the year 1118 (Strik Zijlstra 1993). The manor house, and the castle were in Holland, whereas the town of Millen was in Germany. The moat surrounding the manor house was ingenuously designed in the 17th Century. Intake water was from the adjacent Red Brook. Periodically the moat could be flushed out through a six-foot diameter pipe and into a canal which was about one kilometer (km) to the west and at a lower elevation.

Jacques joined some of his friends skating on the moat. Then he noticed a group of tall black-uniformed Hitler youth muscling the Dutch kids off the ice although there was enough ice for both groups. Jacques decided to go for help from some of the larger Dutch boys, but when he went to the bank to take off his skates he was hit from behind. He swung his skates by the string striking the face of one of the Germans with the sharp blade. He then ran for help. A small army of Dutch people was mobilized and they quickly drove the Germans back across the border. Jacques lived at #14 St. Paulusstraat in Overhoven, barely 1 km from the castle (Briels 1993). Millen Castle and the adjacent manor house were called "de Haan van Millen" in 1939. The de Haan family owned the property since 1823. Jacques Briels, and other visitors to the castle, considered it a "secretive, mysterious and foreboding place. In the many visits Jacques made he never saw any occupants and sometimes left, with the feeling that someone was watching him." (Briels/ van den Berk, 1993).

Jacques' encounter with the Hitler youth is relevant to our main story in Chapter 3. Jacques, like most other Dutch, had never crossed over the border into Germany. Also, it illustrates the non-fraternization between Dutch and Germans who lived close by the border. Jacques was one of the very few Dutch, in that region, who was fluent in German.

In 1939 there were nearly nine million Hitler Youth, aged between ten and eighteen. Each year they swore an oath of allegiance to help Hitler (Koch 1975). Catholic and Protestant youth groups were outlawed. Jacques encountered a few of the Hitler youth at Millen Castle. Five years later, our battalion encountered the Hitler Youth Division at the Falaise-Argentan pocket where only

600 of the entire division escaped (Koch 1975, p. 247). Even German Field Marshall von Rundstedt was appalled at this sacrifice of sixteen- and seventeen-year old German boys in a hopeless situation (Koch 1957; p. 247). German boys and girls manning anti-aircraft weapons were even younger than sixteen. We were destined to meet a few of them at Nuremberg.

Headquarters Troop, 103rd Cavalry (horse), P.N.G.

I was the third of four sons of Anna and Arthur Armstrong. I had settled on a forestry career by 1934. I was partially motivated by President Roosevelt who regularly referred to himself as a forester (Armstrong and Oates 1992; p. 34). However, the darkening war clouds caused a career digression. I was irresistably drawn to following the darkening war clouds whilst my associates were more interested in ball games and big-name bands. I also had an avid interest in horsemanship. The combination led to my joining Headquarters Troop of the 103rd Cavalry of the Pennsylvania National Guard in my closing days of high school (June 1939). When my parents learned of my enlistment they were pleased because they felt war was inevitable and the training I would receive would further survival chances. My two older brothers joined the same troop at a later date.

Our 1939 Army led by 65 officers of general grade

Our 1939 Army was small. Eighteen other countries had larger armies (Carroll p.137). There were 167,000 regular army soldiers; 250,000 National Guard; and 60,000 Army Reservists. There were so few general officers in the 1939 army (65 general officers) that new recruits were expected to memorize names and assignments of the top ranking. This was a ratio of one general officer for each 2,580 enlisted men. The National Guard ratio was nearly identical.

By way of comparison, the 1987 Red Army of the former USSR had one general officer for each 530 men (The Economist, Aug. 28, 1993, p. 18).

Selective Service — The Draft

The Selective Service Act of 1940 was authority for 900,000 men, between the ages of 21 and 35, to be drafted. The 6,500 local selective service boards were assigned quotas and a national lottery decided who was in the Army. The first selectees were inducted in November 1940. By and large, they were not young men. None were teenagers. Lotteries of draft numbers were adhered to at first, but eventually that became subjugated to the matter of deferments. Nearly 12,000 persons were imprisoned for resisting (Flynn 1993: 173), and of these, four thousand were Jehovah's Witnesses. Deferments were for those working in defense plants, or on farms, or those who were fathers (in the early stages). Late in 1942, nearly one year after the United States declared war, Congress lowered the draft-age to 18. Twenty-one years was the drinking, voting, and legal age of maturity.

By the end of the war, Selective Service had files on fifty million American males from age 18 through 64. Men already in the service, such as myself, were not included. The system examined 20 million men and fed 11 million into the services. The peak strength of the Army (1945) was 8,300,000. The total number of people who were in the Army during World War II was more than ten million but there were combat and other losses along the way. Thus, more than 95% of the enlisted men in America's WWII army were draftees from the September 1940 Selective Service Act.

Young men, who had attained the age of 18, could volunteer for our armed forces. Seventeen-year olds were admitted with parental consent. These boys, younger than 21, were not subject to conscription until late in 1942. Thousands did volunteer. At least one, Jack Minton, joined my troop at the age of 14. The Army weeded out these under-age soldiers in 1941. Former President George Bush was flying with the Navy at age 17. He was reputed to be the youngest Navy pilot. Many superb fighter pilots, with the U.S. Navy and the U.S. Army Air Corps, were teenagers. In 1941, the U.S. Army Air Corps was redesignated the U.S. Army Air Force.

The author doesn't intend to slight American women. Back as far as 1939 there were 672 Army nurses. Later in the war the Woman's Army Auxiliary Corps (WAAC) played a vital part, but that activity is beyond the scope of my expertise. Millions of American women worked in war industry.

Teen-age soldiers

Some highly motivated teenagers and young men joined the Canadian Forces in order to become involved in fighting the Germans long before the United States declared war. Most of these were eventually transferred to U.S. forces.

While I was at officer candidate school (OCS) in 1942 I noted younger candidates generally performed better. The OCS drop-out rate, and the set-back rate, were high and generally seemed to hit older candidates. A substantial number of teenagers were commissioned by the President to be officers in the Army. World War II has been referred to as "The last children's crusade".

This does not imply younger soldiers always perform better. It was possibly a convergence of several facets, which may not have been adequately researched. Perhaps men born from 1921 through 1924 suffered greater adversity from the economic depression, or possibly they were more highly motivated by President Roosevelt, or by German ruthlessness.

The Third Battle of Bull Run

A few weeks after my 1939 enlistment I was on manuevers at Bull Run, Virginia, with both Regular Army and National Guard horse cavalry. It was quite an experience being the youngest recruit in the troop. Three older troopers were superb in proffering fatherly advice. This assistance shaped my career. I am indebted to all of them. The drill-sergeant syndrome did not exist, in at least this one outfit.

Much of our equipment was carryover from World War I or before. The only tanks on the Manassas Maneuvers were Six-Ton M-1917 light tanks which were little threat except when two of them unknowingly snared a barbed wire fence and drove down through an infantry formation. American tanks were all less than 15 tons because that was the maximum capacity of our pontoon bridges. The Germans had tanks on the drawing boards which ranged up to 55 tons.

Our Civil-War-model kitchen equipment was built around a blackened sectionalized sheet of iron which was placed over a campfire. Helmets were the WWI Doughboy type. Our vehicles were two-wheel-drive civilian trucks painted olive drab. Weapons were thirty or more years old. The $21 monthly pay was simply shifting the tax burden to a different generation.

The Third Battle of Bull Run, as these maneuvers frequently were termed, occurred but a few weeks before Friday September 1, 1939 when Germany marched on Poland. That weekend the entire world tried vainly to find out what was happening in Poland. Was it only an isolated Polish attack on a German radio station, or was it war? By Monday morning the World knew it was war. The new German government had no integrity at all.

The term "G.I."

I was serious and dedicated, characteristics which resulted in my being called G.I. (government issue) by Sergeant Harley, our supply sergeant, while on the 1939 maneuvers. The term had recently infiltrated into the entire army. G.I., and a few other terms, were informally communicated to all troops with a rapidity which defied comprehension. The term "GI" gradually encompassed all American soldiers, whereas originally it was used for those soldiers who could well have been issued by the government, had the government had the ability to manufacture soldiers as it did manufacture millions of standard issue items. Perhaps most all American soldiers took on the qualities of government issue as they became more aware of developments in Europe.

The instant making of a man

Captain Charles T. Cabrera talked with the troop at the ensuing Monday-night drill. He was certain that the United States would eventually be drawn into the war. Then he stated "Most of you will become officers of a greatly expanded army". It was like a bolt of lightning out of the blue. I became a man in that instant, realizing the responsibilities that would confront me, and also the opportunities to retaliate against the swaggering, bull-necked, be-medalled Germans who had marched so many children off to death camps. Forestry career aspirations were put into a vault.

Cabrera's prediction was right. About half the troopers in that formation eventually attained a commission. A few went to flight school, a few received direct commissions, and the vast majority graduated from Officer Candidate School (OCS).

Mainstream American thought

Cabrera's thoughts were not Mainstream-America. Most Americans wanted to stay out of the European War. Nevertheless, the President and many members of Congress recognized the inevitability. There were influential Americans, British, and French who, at least up to this point, felt that Hitler was an honorable person.

> The Nazi regime wallowed in blood and filth: but the British Ambassador was smitten with the sanguine personality of Field Marshal Goering and with the sophisticated charm of Dr. Paul Goebells. The Teutonic hooligans burned the masterpieces of world literature on public squares in the Reich and persecuted the outstanding German writers: but the great Norwegian novelist, Knut Hamson, and the American jester, Ezra Pound, thought Hitler an eminent promoter, not only of "kultur", but also of culture at large. (Klaus Mann. 1984. *The Turning Point,* p. 268)

Although the majority of Americans were unequivocally against becoming involved in the European war there were many who were anticipating the worst and preparing their families accordingly. The father of Robert Pensock of Hazleton, Pennsylvania had been in World War I. In 1928 he opened a radio store. The family lived in an apartment above the store. He taught Robert the Morse Code and then spent an hour every evening sending dots and dashes back and forth. Robert's "head start" program also included radio repair. The radio training paid dividends and within a few years Robert was one of our Headquarters Battery radio operators.

Fortunately, President Roosevelt understood the true character of Hitler, although he was taken in by Mussolini up until June 1940 when Italy joined Germany in attacking France. German forces invaded Holland, Belgium, and France on May 10th 1940.

> Like virtually everyone else, he (President Roosevelt) overestimated the will of France. the strength of Britain, the willingness of Mussolini to defy Hitler (Steel 1980, p.327).

The Dutch-German border of May 1940

Jacques Briels' father was a leader in the Dutch Young Guards. Dutch soldiers had been preparing for the German invasion for more than a year. Concrete bunkers and road blocks were becoming evident because the route from Millen to Maastricht seemed to be an obvious path for the Germans. Jacques, aged 14, had repeatedly met a German Captain on the border who was intrigued by Jacques' fluency in German. Jacques asked the Captain if he thought the Germans would strike west. His answer was "Yes, and soon" (Briels, 1993). In response to a query about how the Germans would cope with Dutch destruction of the bridge over Red Brook, the Captain responded "The new bridge is already built and is ready to be rolled into place" (Briels, 1993). The German Captain told Jacques that they would go clear to the North Sea in their invasion. At four o'clock in the morning of May 10th, 1940, Jacques was awakened by an explosion followed by machine gun fire. That morning he witnessed the German onslaught. There were the German assault troops, the Dutch resistance from their pillboxes and other defense lines, the Dutch prisoners of war being marched into Germany, and then the endless line of the German 4th Army under von Kluge heading towards Maastricht. Included were thousands of horse-drawn wagons. The 4th Army was under Army Group A, commanded by von Rundstedt (Piekalkiewicz 1980;p. 29).

Within a few days, long lines of Belgian prisoners of war were marched along the same route, but in the opposite direction. Traffic also continued from

Germany because this was one of the main supply routes for the German army. A few weeks later Jacques witnessed the return of the Dutch prisoners of war from their German camps. I believe the Dutch were the only nation whose prisoners were released. Millen castle witnessed it all, but played no part in this phase of the war.

The German main effort, in 1940, was through southern Holland and central Belgium (Willmott 1989: p. 81).

Winston Churchill foresees American intervention

"In God's good time, the new world, with all its power and might, steps forth to the rescue and liberation of the old." (Churchill's "We shall never surrender" speech, 4 June 1940) (Deighton 1993:p. 202)

Meanwhile, back in the U.S.

the U.S. defence establishment, and particularly the U.S. Navy, emerged as the real winner of the 1940 campaign in Europe (Willmott, 1989: p. 101).

Prior to the downfall of France, our troop had met on Monday evenings at the 32nd Street and Lancaster Avenue armory in Philadelphia. The armory included stables and an arena with tanbark floor. We also had two weeks of field maneuvers in the summer. I received one day's pay for each meeting ($0.70) while I was a recruit.

After the defeat of France in 1940, national guard training intensified with numerous week-end manuevers in addition to the weekly meetings. Our weekend training was at the the troop farm. For several years members of the troop had donated their earnings into a fund which leased a farm, paddocks, and polo field adjacent to the old WCAU radio tower in Newtown Square, west of Philadelphia. Instead of buying cavalry mounts, we had procured polo ponies (essentially horses) and, with the old farm house being a club house with bar and dining facilities, had many memorable weekends until June 1940. The regimental Sergeant Major, Herman Ferry, and others veritably lived on the farm. General Hugh Drum, Commanding First Army, played polo there on numerous Sundays. He was a big man and required a new polo pony for each chukker (period of play).

The troop made good use of our mounted pistol range and rifle range. The challenge on the mounted pistol range was to ride along a row of targets at a gallop shooting into each target and at the end of the line insert a new clip into the pistol and shoot the same targets on the way back. One unfortunate soldier had difficulty changing his clip and shot his own horse in the neck. The 45 caliber pistol may have been 29 years old, but it was deadly to man and beast.

Horse Cavalry maneuvers in upper New York State

The entire 52nd Cavalry Brigade, under command of General John Stackpole, was on maneuvers in northern New York State in August 1940. My troop had become Signal Troop of the 52nd Cavalry Brigade. New telephones and switchboards were on hand. Weaponry was still obsolete. Radios were bulky and severely range-limited. Individual equipment was in the saddest state of inadequacy. The Philadelphia Inquirer (8 August 1940), in writing about the New York State war games featured an aerial photograph of the administrative encampment showing the picket lines with thousands of tethered horses. More meaningful to our story was their description of the lack of equipment.

> Some of the soldiers had to use nothing more deadly than wooden guns. Imitation 60 and 81 mm. mortars, carved from wood, formed the only equipment of that type furnished the 11th Infantry Regiment, Philadelphia's own.

President Franklin D. Roosevelt reviewed our horse-mounted brigade from his limousine on August 17, 1940. This formation was probably the last time a full brigade of U.S. horse cavalry assembled. Photographs of the event were featured in numerous Pennsylvania newspapers. However, the war in Poland had proven that horse cavalry was obsolete. Horses could not withstand airplane strafing and Stuka dive bombing. Our Cavalry didn't give up easily. Two years later there was still at least one regiment of horse drawn artillery and two regiments of cavalry. The slowness to adapt to the new warfare was probably linked to inadequate modern equipment.

The slow demise of horse cavalry

European armies continued with horse-cavalry forces and horse-drawn vehicles. In August 1940 the Polish Army had 20,000 mounted cavalry troops. Germany had one brigade of horse cavalry (Piekalkiewicz 1980; p.7). In 1942, the German Army on the Russian front organized the 8th SS Cavalry Division (Stein 1966; p. 202).

By late 1940 this one National Guard unit had been converted from Headquarters Troop of the 103rd Cavalry Regiment to Signal Troop of the 52nd Cavalry Brigade, then to Headquarters Battery of the 2nd Battalion of the 166th Field Artillery Regiment, and then to Headquarters Battery of the 73rd Field Artillery Brigade. These changes were primarily a result of the demise of cavalry and reorganization of divisions from the square division to the triangular division. The troop also had to unlearn the dismounted cavalry drill ("right by fours") and learn standard dismounted drill. There also was the new phonetic alphabet to learn. Prior to 1941 the Army had theirs, and the Navy had a different one. The eleven new General Orders for guards on duty eliminated the twelfth which might not have been pertinent to the new era — "To allow no one to commit a nuisance on or near my post".

The Germans protect Holland from the English?

By late 1940 Dutch people realized the implications of German occupation and plundering. The Austrian Nazi lawyer, Arthur Seyss-Inquart, was the Reichskommissar for the Netherlands. It was a brutal regime and Sadistic Arthur was slated for hanging in Nuremberg after the war. Every town had a German commander. No more than five people could walk or talk together. The Dutch were forbidden to harbor Jews, who were readily indentified by the yellow star which they were required to wear. Eventually, Dutch Jews were shipped off to German concentration camps through the infamous Camp Westerbork (Boas 1985). Eighteen-year old Dutch boys were required to serve one year with the German labor service (Arbeitdienst). Dutch resistance to German rule was dealt with by killing innocent Dutch hostages.

> (T)he Netherlands was treated more severely than its (western) neighbors in many respects: it was ruled by the SS instead of the Wehrmacht, a higher percentage of Jews were deported, Rotterdam became a symbol of massive destruction in Europe and so on. Moreover, for most of the country the war or occupation lasted longer than it did in Belgium and France. (Brands 1970: 691).

The Mayor of Nieuwstadt, and the owner/occupant of the Manor House at Millen Castle was a Mr. Haan. He resigned as Mayor because of disagreement with the German occupying forces. The Germans replaced him with a sympathetic member of the "Nationale Sosialistische Beweging". These collaborators with the Germans were considered traitors and, in September 1944 had to flee into Germany.

National Guard transition to Regular Army

The President signed Executive Order No. 8618 on December 23rd 1940 ordering certain national guard units into the active military service of the United States. There was swift action during that 1940 Christmas season. The January 13th Philadelphia Evening Bulletin spoke of the "transition from a state guard unit to a regular Army unit". The same day Philadelphia's vanguard, of what was to eventually become tens of thousands of young men, was inducted into the regular Army. This was far more meaningful than a national guard unit being called to active duty.

As was the custom, my parents displayed a flag with two white stars in the front window denoting two family members in the service. A gold star would have indicated that one member of the family had been killed in action. Rest assured, there was no heckling of any type by protestors to the military buildup. They had their rights to oppose American military intervention, but they also were decent Americans.

The community (voting district) erected a monument at the old Revolutionary War Church, at the corner of Washington Lane and Germantown Avenue, citing the names of soldiers in service. The monument was not defaced during the war years, at least. Our Church posted a bronze plaque listing persons in the Armed Forces.

Camp Shelby, Mississippi

Our Battery was dispatched to Camp Shelby, Mississippi. The mid-winter trip south, in unit vehicles and Harley-Davidsons with side cars, was superbly planned. The first night was in Gettysburg Armory along with a hot dinner and breakfast. The convoy arrived at Bristol, on the Virginia-Tennessee line, in late afternoon of the second day. After a restaurant-dinner our troopers found beer on the Tennessee side of main street. Legally, the drinking age was 21, but not enforced for men in uniform. The state of Virginia was dry at that time meaning that the state had laws forbidding the sale of alcoholic beverages. This was also true of Mississippi and some other states.

These were the days before Interstate Highways so the convoy meandered through numerous cities and towns. The McManus brothers, Sergeant McGinniss, and others blocked traffic at main intersections with their motorcycles and then raced ahead of the convoy to block the next intersection. Traffic was responsive to instruction from the soldiers. The troop was well received by town's people. Rest stops were along shoulders of rural roads where our soldiers responded to nature's call.

Camp Shelby was one of the largest Army training centers, with an eventual population of 86,000 soldiers. It was located just outside of Hattiesburg, Mississippi. There were the 37th and 38th Infantry Divisions (from Kentucky, Ohio, Indiana, and West Virginia) and the 73rd Field Artillery Brigade in between the two. The Brigade included the 166th Field Artillery Regiment from Pennsylvania (with 155 mm howitzers), the 190th Regiment from upstate Pennsylvania (with 155 mm long cannons), and the 241st Regiment from New Orleans with 155mm howitzers). The 155mm long cannons were a carryover from World War I, where they had been known as "General Pershing's Favorite" (GPF). They had far greater range than the 155 mm howitzers, but also far greater dispersion when the rounds landed. The prime movers for the GPF's were slow moving tractors, probably a variation of farm tractors.

Camp Shelby billets were 6-man pyramidal tents with wooden floors and coal-burning, Civil-War era, Sibley stoves. Wooden buildings were constructed for mess halls, headquarters buildings, latrines, and day-rooms. Design criteria for wooden buildings was a five-year life. Most lasted far more than five years, some still being in use in the 1990's.

There were major reorganization changes when the Ordnance Corps was given responsibility for vehicles. Formerly the Quartermaster Corps had this responsibility. And, the Quartermaster Corps also lost responsibility of cantonement construction to the Corps of Engineers. Each of these many changes were indicative of the lack of our Army's preparation for war. They should have been made several years earlier.

World War II fought with forest products

At the end of the war the Forest Service reported that 48 billion board feet of lumber had been used during the war years for cantonements, hospitals, warehouses, factory buildings, and shipyards. Fifty-three additional billions of board feet of lumber were used for packing, crating, truck bodies, PT boats, and ship construction. The total was equivalent to 20 million homes, or 10 million acres of timberland. Fortunately, there were immense stocks of air-dried softwood lumber available in the late 1930's. Some of these stocks were a result of salvage cuttings of thousands of acres of east coast timberlands which had been levelled by the 1938 hurricane. New England had been especially hard hit. Camp Edwards, at the edge of Cape Cod, was constructed from the finest eastern white pine. This lumber was from the hurricane felled timber, which had not been ruined by the winds shaking trees so hard their annual rings separated. About one-third of the felled trees were ruined by wind-caused ring shake. This always is a major problem with hurricane-felled timber.

There were many specialty uses for forest products. Foresters scoured South America for balsa wood which was needed for life rafts and floats. Biologists scoured South America for quinine trees, the bark being used for anti-malarial medicine. Mahogany was needed for PT boats. Airplane propellers were made from New England yellow birch in the early days of the war. Cork was made from Douglas fir bark rather than the usual Mediterranean cork oak. Wilson Compton, of the National Lumber Manufacturers, stated "Wood has become the great substitute for substitutes." Softwood lumber was not available to the public.

Smoke

I never smoked tobacco although my usual environment was clouded with second-hand smoke. Cigarettes were so inexpensive that it really was encouragement to take up the habit. Later, in Europe, cigarettes were free, usually a part of the rations. Even without the cigarettes, Camp Shelby was densely smogged in on winter days due to thousands of soft-coal burning stoves. Most everyone knew that cigarettes were bad for people eventually leading to chronic "smokers cough". One old West Virginian referred to cigarettes as "coffin nails". Scientific proof of the harm from cigarettes was years away.

Army life in 1941

The shortage of good cooks was probably best illustrated when Bill Green, one of our assigned cooks, served up baked beans, never realizing they should have been soaked in water before baking. I was a good cook, but I never let that become known. Cooking was a dead-end career, frequently resulting in addiction to lemon extract. Why the Army ever issued such large bottles of this alcohol fortified flavoring, no one seemed to know.

Our bugler, Ziegried Mark, later to become a university professor, learned punctuality. From Reveille to Taps Ziegy was there when it was his time. We could have set our watches by Ziegy, if we had had them.

Brigadier General William C. March was Brigade Commander. He requested that I be his orderly, meaning to live in his small house, shine his boots, mix the cocktails, and other chores. I resisted for I had not joined the Army to shine boots and mix cocktails. Headquarters Battery Commander, Captain Edward Brunner, was not happy at my declination. There were repercussions. Kitchen police (KP) for a full week in that era was more than physically taxing. The fifteen-hour days of grueling work left a soldier in such a physical state that his pallor clearly evidenced his recent assignment. Eventually Hugh Gannon, our only college graduate, took the General's orderly assignment.

Shortly after our arrival in Camp Shelby, the battery was assembled in the mess hall where our first intelligence test was administered. Some of the older sergeants rebelled at counting blocks and other facets of the examination. A few weeks later the scores, chronologically arranged, were posted on the company bulletin board. This may have been because the battery commander did rather well. Our high man was private Richard Clayton with a 155. He later went to the Finance Officer Candidate School. Sergeant McGinnis, a real tough Irishman, was second. He was a colonel at War's end. I was number three on the list. None of the humiliated low-score troopers ever complained to the Inspector General, probably because they never knew there was one.

Weekends were three-day affairs before Pearl Harbor. Battery trucks transported soldiers to Lake Ponchetrain military depot and then there were civilian buses into New Orleans. Five percent of the officers and enlisted men had their own cars. The new Jung Hotel (no air conditioning in those days), the Blue Room of the Roosevelt Hotel, and LaLune in the French Quarter became battery hangouts. Wearing of uniforms was required. Sometimes soldiers found that unknown diners had paid for their meal.

Obsolete weaponry

Training included field marches, field communications, and weapons firing. Unanticipated events included a heavy winter rain which flooded the supply tent where weapons were stored. I was selected to detail strip and clean two hundred 45 Colt automatic pistols. These were the prime weapon for the entire battery. Rifles just were not available. On the first Louisiana maneuver some soldiers were required to carve their own simulated rifle from tree branches. If they were not sufficiently realistic they kept at it. British 1917 Enfield rifles were issued before the next maneuver, but the bolt action required the left hand crossing over the rifle. The U.S. Army 1903 rifle was really better. The Browning automatic rifle had less than half the firepower of the German MG-42's. In 1942 the Germans introduced the *Sturmgewehr* the precursors of our current assault rifles. One year later the Russians brought out the AK47 series of assault rifles. The U.S. had nothing to rival these (Perkins, 1987. p.195).

The Garand or M-1 semiautomatic rifle was first issued in 1936 but it was not widely available until 1943. It was a good rifle and even possible to maintain and use in muddy conditions. The history of the Garand reflects most favorably on American ingenuity.

No disloyalty

My brother, Herbert, had attained the rank of sergeant and worked with the brigade intelligence officer. Herbert also had counter-intelligence work to ferret out any disloyalty within the troop. There was background for this unusual operation. During the 1930's, certain ethnic groups lived in their own neighborhoods. Some Philadelphia Italians were supportive of Mussolini and his conquests. Grand block parties were held following certain Italian victories in Ethiopia. Italian flags were flown, street dancing to accordion music, along with fine Italian food, reigned. People with Germanic background were generally widely scattered through the population at large, although Germantown (part of Philadelphia) attracted more. The German-American Bund was a fraternal organization which feted gymnastics. Numerous non-Germanic youth joined the Bund for the benefit of their own gymnastics training. Bund meetings feted the Nazi flag along with the devil's song "Germany Over All". The Federal Bureau of Investigation had records on people who participated in these Italian and German groups. The Army was simply following up on what the FBI had started. However, even before Pearl Harbor it was clear that most of these persons had prime allegiance to the United States. The only known German spies who landed in the United States by submarine (Douglastown, Long Island) were initially hosted by a physician of German descent. The ensuing trial revealed no prior arrangement for this hospitality. The spies were executed (Medina 1975).

Louisiana maneuvers

The Battery participated in Louisiana maneuvers in the summer of 1941. General March was relieved of command and returned to civilian life. Colonel Bullard took over the Brigade. He made many changes, especially in the luxurious life-style the officers had been living. Throughout the country, numerous senior national guard officers were relieved of duty and returned home as civilians whenever it was apparent that training had been neglected. This was the advantage to the Army of federally activating the National Guard as opposed to calling them to active duty.

There were numerous deaths from motorcycle accidents on 1941 Louisiana maneuvers. These led to the decision not to use motorcycles in the Army, but to rely on jeeps. Motorcycles were used by other armies, but I don't believe we used them after 1941.

Light aircraft (Piper Cubs) were occasionally employed to verify the camouflage training. If units down below were detected, they were bombed with sacks of flour. Some troops retaliated with rocks. Then followed a written order forbidding the throwing of rocks at airplanes. Americans are noted for their throwing ability. In later years I met one West Virginia youth who could stun a squirrel with a rock.

In October 1941, as the Louisiana maneuvers ended, nearly half the battery was assigned to temporary duty with the 2nd Provisional Antitank Group for maneuvers in North Carolina. They returned to the Battery at the end of this experiment. This was one of the many trial organizations which eventually concluded in tank destroyer battalions. Tank destroyers were lightly armored tracked vehicles which supposedly were more agile than tanks. The experimentation was in response to the 1940 German "blitz" warfare in France.

In Louisiana, and eastern Texas, there were thousands of acres of cut-over forest with tree stumps as far as the eye could see. This was ideal for mechanized military maneuvers. The same terrain is no longer suitable for such maneuvers. Today, these lands are a green ocean of loblolly pine trees, entirely too dense for tanks and other military vehicles. Heavy use of timber during the war, coupled with anticipated pent-up demand for wood after the war, and the 1943 favorable long-term capital gains treatment of timberland earnings, all motivated land owners to commence planting pine seedlings and managing their forests. Those trees, planted in the 1940's, are now sawtimber sized. In the last decade industry, and other forest owners, have planted more trees each year than the CCC did in its nine-year existence. Today, however, the trees are machine planted. A second golden age of forestry has swept the southland in the early 1990's as a result of the closing down of much timbering on national forests of the Pacific Northwest. Lumber prices doubled in the year 1993.

Troops on the 1940 and 1941 Louisiana maneuvers probably best remember the armadillos, poisonous coral snakes, and billions of chiggers (a small insect which bites humans leaving a two-week swelling). Chigger attacks on troops sleeping on the ground, following a rainstorm, were so severe that some units were virtually disabled. There were no insecticides at that time except possibly powdered sulphur. Our Civil War all-time record-setting non-battle casualties (244,771 for the North) were partially attributed to the northern soldiers being unaccustomed to the heat, humidity, insects, and poison ivy of the Southlands. (Note: the military use of "casualty" is not confined to killed in action. The term includes personnel evacuated, hospitalized, and also those killed in action.)

Japan and Germany declare war on the United States

On December 7, 1941, I was the day-room orderly running the battery-sponsored crap game and black-jack game. Gambling profits were high, but they all went into a company fund to sponsor beer parties. That Sunday the radio music was interupted for the announcement of the attack on Pearl Harbor. The world would never be the same. Instantly, President Roosevelt had the backing of most Americans. Overnight, most every American was involved indirectly, or directly, in the war.

Our Naval losses at Pearl Harbor were not publicly disclosed, and although reporters could easily have collected information from the witnesses to the Japanese bombing, and then disclosed them, they did not do so. Slowly, word leaked out, but it was too incredible for belief.

Japanese population was 95 million which was greater than Germany's 85 million. United States' population was 140 million (Willmott 1989:p167).

Training in 1942 with obsolete weaponry

Training intensified, and in early 1942 the Brigade was back to Louisiana for maneuvers. Even at this late date equipment shortages plagued the Army. Trucks, with a sign reading "Tank" along with a log protruding over the truck cab were simulated tanks. Anti-tank guns and even rifles were simulated. Four-wheel drive trucks were not available, except for the new Jeeps which were starting to arrive. The word "Jeep" came from Al Capp's cartoons combined with an abbreviation for "General purpose equipment". Immediately after 1942 Louisiana maneuvers, the Brigade was re-assigned to Camp Sutton staging area near Monroe, North Carolina, where Brigadier General John Lewis assumed command. The 190th Field Artillery Regiment, with their 155mm long-rifle cannons, was alerted for shipment to North Africa. The remainder of the Brigade remained in the U.S. for at least a few months.

My older brother, Herbert, went to the Cavalry Officer Candidate School at Fort Riley, Kansas. He later served in Europe with General Patton's Third Army. Our paths never crossed in the European war.

Artillery School Officer Candidates

A month later, I applied for officer candidate school. My battery commander, Captain Brunner, strongly opposed commissioning any one as young as I. Nevertheless, I became the youngest soldier to appear before the Brigade OCS Board. The Board strongly recommended approval for the Artillery

School. In view of the controversey, Brigadier General John C. Lewis personally interviewed me, and then decided I should attend Artillery OCS.

I travelled by train to the Artillery School at Fort Sill, Oklahoma. Railroad travel during the war was not pleasant. Much of the rail equipment was being used for troop transport. Manufacture of new rail-equipment was subjugated to the manufacture of army tanks and trucks. Retired rail cars were recalled. Some of the ancient cars had pot-bellied stoves for heat. Trains were crowded with soldiers on leave, families trying to visit husbands and fathers, and defense workers moving to new jobs. Soldiers felt compelled to give up their seats to mothers with children and then sit on their suitcases for the remainder of the trip.

The most important challenge for officer candidates at the Artillery School were ten firing problems. Classes were taken to an observation post and assigned a target some 3,000 or more yards away. Then a candidate was called and was expected to come out of the bleachers shouting his commands to the telephone operator for transmission to the howitzers. He then had to make adjustments, while the instructor was shouting at him to do something. The instructor's antics were intended to simulate the pressures of combat. Older candidates frequently became un-nerved by the instructor. Youngsters, like myself, frequently did very well.

We were class #43 of the Artillery Officer Candidate School. Initially, the 440 candidates were assigned to five-man huts alphabetically. My hut included John Alfred Ames, John Ellsworth Andersen, Peter Richard Andre, John Kenneth Stuart Arthur.

The five topics at OCS were gunnery, motors, tactics, communications, and logistics, which were taught in block instruction. That is, we would have ten straight days of communications and then go on to a new topic. Strangely, each one of us had background in one of the five topics. Hence, we cooperated and encouraged each other. Ours was the only hut, of about 85, which graduated intact. Three hundred and thirty-two lieutenants were commissioned in December 1942. Some of these had been set back from earlier classes. "Commissioned officers" are those who have been commissioned by the President.

I was assigned to a regiment in the same brigade I had been with as an enlisted man in accord with the request of the Brigade Commander. This was highly unusual. Coincidentally, the regiment had just been temporarily assigned to Fort Sill where they had ample ammunition in the role of supporting Artillery School training. The last remaining regiment of horse-drawn 75mm howitzers was still active at Fort Sill.

Newly commissioned lieutenants were granted two weeks leave. I was home in Germantown, Philadelphia, for Christmas of 1942. Neighbors, friends, and especially school teachers were delighted that a young man from such a poor family had been commissioned. My high school classmates had just become eligible for the draft. Many sought advice.

Platoon Leader

Upon completion of leave, I reported to my new regiment at Fort Sill, Oklahoma. I was assigned to the 2nd Battalion of the 166th Field Artillery Regiment, commanding the anti-tank platoon. The prime weapon of the anti-tank platoon was a new model of the 37mm anti-tank cannon. It was already obsolete, totally incapable of penetrating the armor of any of the new German tanks. It did have a marvellous sighting system, but that was little use if the rounds could not penetrate armor. Despite my youth, I was well received by the men in the platoon. The platoon sergeant, George Mertz, has maintained contact for at least the next fifty years, and was a consultant in the writing of this book.

The battalion commander was a Lieutenant Colonel Bailey, a reserve officer who normally was a Philadelphia lawyer. Officer meals, in a pyramidal tent, were a most pleasant affair. There was no hazing of second lieutenants. Senior officers were addressed by title, such as "colonel". The word "sir" was not used.

The battalion lived in tents on a wind-swept Oklahoma prairie. The wind did not permit stoves in the tents assigned to officers because those tents were so small the stove would have to be close to the canvas. January was cold, but I had four wool blankets in my bed roll and slept soundly fully clothed. Tenting was becoming normal, and with few exceptions it was the mode for five straight years. Most World War II officers pitched their own small pup tent when they were not in foxholes.

In February 1943 the Battalion was assigned to Camp Blanding, Florida. The well-planned camp was built around a clear sand-bottomed lake. Hiking in the loose sand was taxing for some of the men.

While we were at Camp Blanding, the Brigade Commander, General Lewis requested each regiment to nominate a lieutenant for his Aide de Camp. I was nominated by the 166th Regiment. In a friendly interview General Lewis was most understanding when I told him I would rather serve with troops. Colonel Boettinger, the regimental commander, was not happy with the declination. I was aware that I had also declined an immediate promotion. Later, I learned that General George Marshall, Chief of Staff, forbade West Point graduates from becoming an aide de camp. He felt they had more important responsibilities.

The next assignment was Camp Gordon, Georgia. There, the main problem was coping with bed bugs. It was strange how docile some of the hard-drinking older men became when they needed a company officer to bail them out of Military Police jurisdiction. I learned there is no such thing as a "tough guy".

The 166th Field Artillery Regiment was equipped with 155mm howitzers. In deference to the aforementioned triangular-division concept the regiment was dissolved and reorganized as separate battalions in March 1943. This transition was quite smooth. The 2nd Battalion Headquarters Battery, with the anti-tank platoon, became Headquarters Battery of the 939th Field Artillery Battalion. The Battalion functioned under a "Group Headquarters".

Tennessee maneuvers in 1943

In early 1943 the Battalion was on Tennessee maneuvers. My anti-tank platoon was cited for outstanding aggressive action, supposedly having knocked out numerous enemy tanks. The maneuver-officiating officers, in giving so much credit to my platoon, were not aware of how misleading their observations really were.

Half way around the world, at about the same time, the Army had just found the 37mm anti-tank guns were worthless in the Battle of Kasserine Pass in Tunisia. The guns could not penetrate any German tanks. Some of the 6,300 American casualties were soldiers manning the "paint scratcher" as the anti-tank gun became known. One hundred and eighty three American Sherman tanks, with 75mm cannon, also were lost at Kasserine Pass. German Mark VI tanks, with their 88mm guns, outclassed our equipment.

In both training, and on maneuvers, I was pleased to find that officers performing well could have freedom of action. The Army tolerated self-reliance, and even encouraged initiative.

Assembly of a new battalion

There was variation in Army training procedures. Generally, "selectees" went from home to a reception center for a few days of testing, physical examinations, outfitting, and personnel record compilation. This was followed by assignment to a replacement training center (RTC)'s. They are now referred to as "boot camp", although I never heard that terminology during the war. This is where they were subdued by their drill sergeant, and also learned to function as a team. Then they were reassigned to the unit which they would serve with for the entire war. After the war, there were studies by the General Accounting office on training sequencing (GAO 1977), but no significant changes have been made.

In March 1941, there were twelve large replacement training centers (RTC). Three of them were artillery centers (Fort Bragg, Camp Roberts, and Fort Sill). Branch immaterial training did not become effective at the RTC's until August 1944. The dual mission of RTC's was to provide fillers for the initial vacancies in units being activated and also to provide combat loss replacements. The immense expansion of the Army, along with activation of so many units, sporadically created a demand for fillers which was greater than the RTC's could supply (Palmer 1947).

Army Ground Forces (AGF) was established in March 1942 under the command of General Leslie McNair. AGF was responsible for the RTC's. General mcNair, a former artillery officer, had excellent judgment, logic, and common sense. Unfortunately, he was killed in Normandy by our own aerial bombing in July 1944. AGF established the mobilization training program (MTP) which included individual training, unit training, and combined-arms training.

Two months after Pearl Harbor the RTC's were not able to keep up with the demand for fillers and so newly activated divisions assumed the role of RTC's for their own fillers. This continued until mid-July 1942 when, ostensibly, all fillers received training at RTC's. However, in July 1943, when the War Department extended individual training from 13 weeks to 17 weeks, the RTC's again became inadequate (Palmer 1947).

New battalions of towed 105mm howitzers were activated in groups. There were 17 battalions carried over from the old army. The seven more activated in February 1943, and two in May 1943, had received conventional fillers from Artillery Replacement Training Centers. Four more towed 105 mm howitzer battalions were slated for activation at Camp Rucker just as the RTC individual training was expanded to 17 weeks. The RTC's could not supply the fillers for these battalions. This created the need for new procedures. Thus, newly inducted men were shipped directly from reception centers to the four battalions being activated at Camp Rucker. The deviation from normal procedures was excused under the title of "experimentation". The four battalions were the 282nd, 283rd, 284th, and the 512th. No more non-divisional light artillery battalions were activated until April 1944, when six were so designated. The method of training fillers for these last six battalions has not been verified.

The unbiased makeup of the experimental battalion

In the case of the 283rd Field Artillery Battalion, reserve officers were assigned to the senior positions. Recent graduates of officer candidate schools were assigned as lieutenants. Most, but not all, of these new lieutenants had some college education. The 939th Field Artillery Battalion sent four officers, including myself, as part of the cadre. Captain David Evans of Allentown, and I, were from Pennsylvania which probably was a fortuitous development. However, it may have been that the 939th had been requested to assign Pennsylvanians.

Fifty-two non-commissioned officers, from the 198th Field Artillery Battalion of the 38th Division, then in Hawaii, completed the cadre. Most of these excellent non-commissioned officers were from Louisville, Kentucky.

Our filler personnel were 323 Pennsylvanians. Precisely one-third of these were from Philadelphia. The remainder were selected from 102 Pennsylvania towns and cities. Generally, only one or two men were selected from each town/city. An amazing 47 of Pennsylvania's 68 counties were represented. The intent of the wide geographic spread was probably to draw an average group of Pennsylvanians so as to minimize bias in the experiment. It also minimized the probability that any town would suffer unduly in the event the battalion became embroiled in a major fire fight. Hazleton, with 17 men, and the small town of Vandergrift in western Pennsylvania with 15 men were notable exceptions. Within the battalion the men were evenly distributed. For example, each of our five batteries received three men from Vandergrift.

It seemed to me that, collectively, these men represented an average Pennsylvania young man. There were a few from very wealthy families and many who came from the poorest families. Normally, some would have gone to college at age 18, but in the World War II draft there were no exemptions for college students, other than that draft boards usually permitted college students to complete their current semester prior to induction. Neither were athletic champions exempt from the draft.

> Lindley V. Righter of Headquarters Battery had signed a baseball contract with the Philadelphia Phillies as he completed high school. He was a pitcher. He spent all of his weekends practicing pitching while at Camp Rucker and Fort Riley. He did not make it into the Pros after the war (Mamlberg 1994).

There were no documented differences in attitude and ability between young men from Pennsylvania and other states or regions. It was noted that New England recruits had a larger percentage of muscular men than other regions (Stouffer 1950).

The Pennsylvanians

The trainload of Pennsylvanians arrived at the Camp Rucker rail-head. They were in remarkably good spirits after a long trip from New Cumberland Barracks Reception Center in Pennsylvania. They had no idea of their destination until they neared Camp Rucker. The mystery was heightened by blackout curtains being drawn on the antiquated, stuffy, coaches whenever they passed through a station. In deference to the dirt and grime the men had changed into fatigues (blue-denim work uniforms) until they neared their destination. While on the train they were introduced to "battery acid" as the canned grapefruit juice was known. Sandwiches were delivered to the train at various stations. Most important, they became acquainted with the guys they would be with for two years.

Milton Anderson, of Lancaster, was unique because he had a prior enlistment. He had undergone basic training with the 82nd Airborne Division and had been on the Invasion of North Africa with the 1st Armored Division. But, most unusual to his story — both he and Lieutenant Colonel McDonald had returned from North Africa at about the same time and he had talked for more than an hour with McDonald while neither of them realized that very soon they would both be in the same battalion. And then, to top all this off Milton met up in France with his buddy who had first joined up with him and the 82nd Airborne. The Army is really a very small place, and not impersonal to the extent believed by many people.

Ethnic mixing

Most of our Pennsylvanians were about nine years old in 1933 when the depression was at its worst. There were others with a background similar to William Davis. No one could be classified as typical because the variance was very large.

> I was born in Beaverbrook, a typical "Patch Town" of about 300 people. It was built by a hard-coal-mine company for their employees. The deep mine and its attendant colliery were close by. When I was three years old we moved to Hazleton where

about 40,000 people from every country in Europe and Great Britain lived with concomitant frictions. I grew up in an Irish neighborhood and had some problems.

Many Patch Towns were the scene of action by the "Mollie Maguires" (a secret Irish organization) seeking better coal miner wages and also trying to overthrow the capitalistic ownership of the mines. The Klu Klux Klan was active in opposition to the Molly Maguires. The continual feuding between ethnic groups, coupled with the deep economic depression, aggravated childhood development.

Our family never had an automobile so the walking and biking put me in such good physical condition that I had an easy time in Army training. Many old stripping holes from the past coal mining were filled with cool clear water. I became a good swimmer. I was well versed in camping and taking care of myself because of three years in the Boy Scouts.

I became acquainted with guns at an early age. When I was age 12, I was old enough to carry a rifle on hunting trips with my dad. When I was 16 I learned to drive a delivery truck. I was one of the few men in Headquarters Battery who knew how to drive.

When we became teenagers we fell away from the animosities of our parents. Our World War II Army dissolved the remnants of active enmity between most ethnic groups (Davis 1993).

Bill Davis was assigned to the Personnel Section of Battalion Headquarters. He worked closely with Walter L. Scott, the Battalion Sergeant Major, and knew all the officers and most of the men in the battalion.

These teenagers had just become subject to the draft and hence some volunteered. I never knew which ones had volunteered. There was no real reason for my knowing. World War II was the second great mixing of America's ethnic groups. The 1930-era Civilian Conservation Corps (CCC) had been the first ethnic mixing. The numerous ethnic groups assigned to man this battalion rapidly learned to get along. Jewish soldiers volunteered to staff the mess hall, and other essential duties, on Christmas day. The names they called each other were a good-natured carryover from the past and not taken seriously. Officers did not try to stop the derogatory name calling because that would have caused trouble tantamount to creating "teacher's pets". The 1990's advocates of "political correctness" would have been aghast. Yet, viewed in retrospect, it reinforced our overall unity. During training, and during combat, there were never any groupings, or gangs. The gun sections lived and travelled together, but sleeping in shifts, along with the requirement to continually man each howitzer, even at meal time, meant their actual associates spanned the entire battery. The ethnic mixing within this one battalion was a grand success. Many of the Philadelphians were from the south end which was then, and probably still is, the tough end of town. The first week's friction between Philadelphians and up-state Pennsylvanians dissolved rapidly. No one could have wished for better success.

One year older than the average

It seemed as though all of my military contacts were Pennsylvanians. The new soldiers and I struck it off very well from the beginning. Possibly they appreciated someone from Philadelphia. Possibly it was because I was just one year older than the average. Eventually, their families looked to my family, in the Germantown section of Philadelphia, for information and solace after casualties occured. One Gold-Star mother travelled a greater distance. After Lieutenant Jack Hare, of Battery B, was killed his Mother, Mrs. Nathaniel J. Perkins of Carysbrook, Virginia, visited my family. Jack's death was traumatic to his family in Virginia as well as to our Battalion Commander.

Probably none of the battalion visualized the rough path which lay ahead, nor the ensuing time frame. By the time the battalion landed in Normandy men and officers knew most every member of the entire battalion. There were no name tags. Possibly name tags were considered a "crutch" which might nullify the need to use one's memory. Platoon leaders were expected to know all of their men by name within a few days of being assigned. By war's end, battalion members were generally closer to each other than to their own siblings. Many have maintained contact with others since the war.

Camp Rucker, Alabama

Camp Rucker did have wooden tar-papered buildings along with a few tents. It was hot. There was no airconditioning. Training was never interupted because of weather conditions. Individual and unit training were integrated. For example, vehicular maintenance, known as "motor stables" were conducted at 4 p.m. daily after the first week. This required that the men receive specific assignments such as gun sections. Some of us may have referred to it as basic training but it was really integrated training. We did not have drill sergeants, other than each platoon had its platoon sergeant. Lieutenants taught the larger classes, usually under pine trees. Incidentally, 1990 research has revealed that pine trees exude a subliminal fragrance which favorably affects human attitude (Armstrong and Oates 1992, p.51). Classes were regularly inspected by the battalion staff and hopefully the instructor knew his subject to the point where no field manuals were in evidence. Training was a six-day week affair along with an occasional Sunday. A few officers and men (Herneisen 1993) had billeted their families in nearby towns. They didn't have much time with their families.

The heavy emphasis on physical conditioning, was highlighted by frequent twenty-five-mile forced marches with full pack. Eventually, this became a dominant test of the battery, administered by officers from group headquarters. There was little, if any, jogging or running.

Most of the men had never driven a vehicle. There were a few Civilian Conservation Corps (CCC) veterans who became driving instructors for our 2 1/2 ton GMC trucks. Lieutenant Colonel Hugh McDonald, the Battalion Commander, issued an order that no one would be promoted to corporal until they had a truck-driving license. That was the only motivation it took for volunteers to stand in a waiting line to learn truck driving in the evenings and most all day on Sundays. The instructors had to be good to teach gear shifting on large trucks to men who had never even driven an automobile.

Fred Kimmich — our elder trooper

Fred was about twice as old as most of us. He had a daughter who was sixteen. He had been in the CCC or the National Guard before being called into the Army. Age didn't matter. He did everything that was required in our training. Camp Rucker was sandy, hot, and dry in August. One day the heat was getting to us. Fred convinced us that he had the solution to our thirst. He gave us some of his chewing tobacco. After that, we never told him how thirsty we were and never again did we chew. Fred slept with his eyes half open. If you told him you had guard duty at 2 a.m. or 4 a.m. he would waken you ten minutes before your tour. He had a brother in Army Ordnance (Parker 1994).

Somewhere in Germany an artillery shell was destined to kill Fred.

The experiment with tracked prime movers

The 283rd Field Artillery Battalion was selected from the four battalions in training to receive an experimental tracked-personnel-vehicle to pull our howitzers. The track vehicles had good capacity for hauling the gunners, their equipment, and a basic load of ammunition. However, the track vehicles were just too much power for the light 105mm howitzers. The howitzers suffered accordingly, and the battalion went back to the 2 1/2 ton GMC trucks for prime movers. This experimentation, at a time when the Nation had been at war for nearly two years, is typical of reasons for American delay in assaulting the European Fortress.

Discipline

Discipline within the battery was excellent. There were no courts martial, no desertions, and few if any AWOL's (absent without leave) while "C" Battery was in training. There was a court martial in "A" Battery in mid-1944. Company punishment was usually a requirement to dig a hole six feet by six feet by six feet in off-duty time, then have it measured by the duty sergeant, and then refill it. There was never a single complaint to the Inspector General, nor congressmen, nor any other official during the battery's entire two and a half year existence.

The Dutch Underground Resistance recruits a new member

Jacques Briels joined the Dutch National Brigade (H.N.B.) in 1942. This was one of the undergound groups dedicated to hiding the Jews, making life miserable for the German occupation forces, and the underground also tried to eliminate some real bad Nazi or traitor apples by taking them off the list of living creatures (Briels 1993).

Meanwhile, at Camp Rucker

A team of officers from Army Ground Forces was charged with testing units after a few months of training. If the unit passed the test it was deemed ready for combat. Otherwise, a new commander was appointed and training was resumed. Artillery testing primarily focused on adjusting fire on designated targets. I was the forward observer for Battery "C", and all went very well. Unfortunately, the "A" Battery observer found an insurmountable problem. There were four howitzers in each battery which when fired in unison should direct their artillery shells in a parallel, and close together path. Sometimes the initial rounds at a target were from only the two center howitzers so as to conserve ammunition. This was the method used in our testing. However, the "A" Battery observer found to his consternation that the requested two rounds were widely divergent. Something was radically wrong at the gun position. One round was probably correct, the other was not. In sensing the round, and making adjustments to hit the target, he guessed wrongly. I never learned what had gone wrong at the gun position. The battery executive officer had not been popular with the men. He was transferred to another battalion after this event. Overall, the battalion passed the test.

Fort Belvoir, Virginia

Fortune smiled on me for the Christmas season of 1943, as well. In early December I was assigned to an artillery officer's camouflage and explosive course at Fort Belvoir, Virginia. The school closed for the Christmas season and re-adjourned in January. Once again I was home for Christmas. In mid-January I returned to Camp Rucker. (Digression: In the 1950 era I was an officer in the Corps of Engineers and had several longer assignments at Fort Belvoir. And, from 1957 until 1962 I was the explosives expert on the Engineer Committee of the Infantry School.)

Tennessee maneuvers in 1944

In early 1944 the Battalion was assigned to the Tennessee Maneuvers. I, having been on Tennessee maneuvers less than a year earlier with another battalion, was dispatched along with a few enlisted men as the advance party. In the meanwhile, the battalion, along with all of our equipment, was loaded onto trains for shipment to the Tennessee maneuver railhead at Shelbyville. Weather was not cooperative and there was a few days delay in the battalion's arrival. I located the initial gun positions near Murfreesboro, and then found warm dry shelter for my small team in a guest home where we were able to conceal our army vehicles in a barn from the roaming military police. February and March in Tennessee were cold, rainy, and muddy. Troops were not permitted to use any shelter except their own small pup-tents. A year later it was generally agreed that Tennessee maneuvers were more disagreeable than actual combat.

At least one soldier was selected to attend umpire school for officiating the maneuver. He was from "A" Battery, and possibly the First Sergeant selected him "to get me out of his hair because I had a few run ins with him. It turned out to be a good deal because I was able to get three-day passes on weekends" (LeVan C. Reber, 1993).

Church services were available in Camp Rucker and also on the Tennessee maneuver. Usually they were well attended. However, there was one cold, rainy Sunday in Tennessee when only the Chaplain and I met under a dense hemlock tree. It was a short service.

Individual combat readiness testing at Fort Riley

All men and officers were required to go through the simulated-combat range and obstacle course. You crawled under barbed wire while live rounds of ammunition were being fired just over you. There were holes which had been prepared in advance with nitrostarch explosives. These were periodically detonated, and then reloaded and wired after each wave of men went through. It rained heavily on our night. The men went through the course and then returned to the barracks for hot showers. Officers were scheduled last. We, in Service Battery, had responsibility for reloading the demolitions and that night we had surplus nitrostarch. Instead of putting one block in each hole we used several and then enjoyed watching our officers get sprayed with mud (Parker 1994).

Fort Riley, Kansas — Initiation into Operation Overlord

The battalion was approved and ready for combat. We left Tennessee by train for Camp Funston, adjunct to Fort Riley, Kansas, arriving on 27 March 1944. We were well located for shipment to either the European or the Asiatic front. Then, on the 6th of June, 1944, the very day of the great Normandy invasion, our battalion was alerted for immediate shipment to Europe.

The American forces' two and a half year delay in engaging German forces in Europe can partially be attributed to the poor state of American weaponry and other equipment at the time the United States declared war. Mobilization of American industry to the point where airplanes, Liberty Ships, tanks, and other equipment were rolling off assembly lines at a phenomenal rate did take time.

Never before in history, and probably never again, will military troops have the luxury of years in training to fight a specific enemy. However, of all of the Americans who fought in Europe, probably only five percent had been trained to the extent I had. All members of the 283rd Field Artillery had at least one year of training. Training does pay! Hopefully, the United States

Army will never again be in a position where Congress has ignored funding for research and development of equipment and weaponry. "The prewar army budgeted only 1.2 percent to research and development of new weapons" (Carroll, p.138).

Chapter 2
THE NORTHERN FRANCE CAMPAIGN

Preparation for overseas service

Our 283rd Field Artillery Battalion was alerted for shipment to Europe on
the very day Allied Forces landed in Normandy, "D" Day, 6 June 1944. Preparation for the move by train from Fort Riley, Kansas, to the New York Port,
and then the ocean voyage to Great Britain, commenced on an around-the-
clock basis. Ocean air and salt water can be ruinous to rifles, howitzers, and
other equipment. Packaging of army materiel for such voyages was well
regulated, requiring large quantities of cosmoline (heavy grease), waterproof
paper, wooden boxes. and inspections at each stage. The United States used
43,000,000,000 board feet of lumber for packing, crating, ammunition boxes,
and vehicular bodies during the war. Cantonment, factory, shipyard, and boat
construction lumber requirements were discussed in Chapter 1. American
packaging methods made great strides in World War II, led by the Engineer
Research and Development Laboratories at Fort Belvoir, Virginia. In later
years one of the engineer leaders at the Lab, Walter Armstrong, told me about
the war time round-the-clock work. Every decision had to have a firm basis.
For example, the number of rolls of concertina barbed wire in a pallet, was
based on the normal width of railroad boxcars.

Unit cohesion

The Fort Riley medical examination caused the last-minute transfer of one
soldier to a stateside unit because of medical reasons. He resisted to no avail.
He wanted above all to remain with his battery in their forthcoming combat
with the Germans. Miraculously, he did arrange for his 1945 discharge to be
from the 283rd Field Artillery Battalion (Ling 1993). This instance typifies the
morale, comraderie, and unified sense of purpose of the battalion. Time and
time again, our wounded all managed to return to the battalion, despite moves
to put them into the replacement system when they healed. The motive for
some wounded declining hospitalization hinged on fear of not returning to their
comrades. Unit cohesion was greater than average because this unit had been
one team since the men entered the Army. They had never been to "boot camp".

> Colonel McDonald did his best to get the hospitalized men back.
> In one case I was assigned to go to a replacement depot and pick
> up a man who had already been assigned to another outfit. We
> brought him back (Davis 1993).

Camp Kilmer, New Jersey

At Fort Riley, vehicles and howitzers were loaded, blocked, and tied onto flat-bed rail cars. Packed kitchen equipment and other supplies went into box cars. Finally, the troops loaded onto sleeping cars, a real luxury at the time. It was a three-day trip to the staging area at Camp Kilmer, New Jersey. Unloading rail cars, and movement into barracks, was swiftly accomplished during hours of darkness. At 5 a.m. that day most personnel were granted a 48-hour pass. This was adequate for the majority, including myself, to pay a final visit to Philadelphia, and other nearby locations. It was obvious to my family that I was headed toward Europe, but the matter was never discussed. Most important, there were no tears nor expressions of concern for my safety while I was present, and especially at my final leave-taking.

> I spent the evening with my father who was an electrician at the
> Philadelphia Ship Yards (Davis 1993).

Our men and officers had scarcely returned from pass when loudspeakers announced the battalion was to immediately move to the New York Port. No one missed the boat! This was one more indicator of morale, determination, and above all a sense of purpose.

Our ship, the Louis Pasteur

The Louis Pasteur had been launched at St. Nazaire, France, in 1938 as a luxury cruise ship. It was 750 feet in length, with a gross tonnage of 30,000 tons, and a top speed of 26 knots. German submarines being used in the Atlantic had top speeds of 16 knots on the surface and 7 knots submerged (Seaton 1981: p. 129).

The French government used the Pasteur for the 1940 dash to Halifax, Nova Scotia, with the gold reserves of the French National Bank. Advancing German forces had an objective of capturing this gold. They failed. The British assumed ownership of the Pasteur to prevent her falling into the hands of the French Vichy government. The Vichy goverment had become dupes of the Germans. In Canada, the Pasteur was stripped of luxury accommodations and converted into Spartan living quarters for 5,000 American troops. Anti-aircraft and anti-submarine weapons bristled on decks. The Pasteur continued to transport American troops for the entire war, with a British crew. She was a Godsend because she could outrun German submarines. The Pasteur travelled alone — not in a convoy. However, she had not been designed for the rough north Atlantic and could develop a considerable roll to the detriment of passengers not immune to sea sickness.

Batteries "A" and "B" of our battalion were selected to man the ship's anti-aircraft and anti-submarine guns. Officer passengers were in above-deck staterooms, but sixteen were assigned triple-bunk beds in what formerly had been single accommodations. Enlisted men were packed as tight as possible below deck where fresh air was minimal. Troops were allowed on deck during certain daylight hours. The ship was completely blacked out at night. Initially, several abandon-ship drills were held. No one needed encouragement to keep their life jacket with them at all times. The fourth day out, the generators stopped and we drifted helplessly in the middle of the Atlantic. The Pasteur was a sitting duck for German submarines. We went into immediate abandon-ship stations. Troops who had been assigned to man anti-aircraft and anti-submarine guns were on full alert. Fortunately, repairs were effected within a few hours and we all breathed more freely. There were several submarine scares. Radio security was good, and we pursued a zig-zag path. We navigated up north of Ireland and then down into Liverpool encountering several floating mines along the way. Some of "C" Battery men, including Myron Brenner, were assigned to "mine-watch duty" out on a precarious station on the bow of the ship. Brenner was from our Hazleton contingent. There were three instances where the ship stopped inorder to ascertain the direction the mines were moving. One mine was detonated by the Pasteur's guns but the Captain was reluctant to make too much noise which could attract submarines. The north seas were rough and most all of us were eventually seasick. There were no available medicines to combat seasickness. One unfortunate soldier, George Glarner, was seasick before the ship had left New York. He was destined for a mind-shattering experience in Germany, from which he fortunately survived.

The Pasteur was returned to the French after the war. She was used to transport the French Foreign Legion for a while. In 1957 she was purchased by a German shipping company, refitted, re-named the Bremen, and used for luxury cruises between Bremen and New York. She sank in a tropical storm in 1981 as she was being towed to the scrap yard. This marvelous ship saved many American lives by her safe wartime transport.

Liverpool, England

The 283rd Field Artillery Battalion arrived safely in Liverpool. It was good to disembark after those rough final days of the voyage. Obviously we were in the war zone. There were numerous bombed out buildings and wharfs. Sunken ships littered the harbor. Dark smoke belched from most every factory. Anti-aircraft weapons were visible most everywhere. Women were doing heavy manual labor. It was a depressing scene especially when we realized similar scenes were to be found in all parts of the world. It was all the result of one nation wanting to rule the world, "Deutschland Uber Alles."

Abergavenny, Wales

Our arrival in Liverpool, England was superbly planned. A host artillery battalion, commanded by one of my former OCS instructors, met us on the dock and then loaded our battalion aboard trains for the trip to Llanover Hall, a large estate near Abergavenny, Wales. We had to drive to Cardiff to pick up our howitzers and trucks. Our camp was a set of Quonset huts set up on the grounds of the estate. Food was scarce throughout Great Britain, and we were usually hungry.

Battalion made arrangements to use the simulated artillery range of the British Artillery Training Center to give additional training in adjusting artillery fire to the senior non-commissioned officers and enlisted men in the battalion who would be serving on forward observer teams. Lieutenant William Grohne was selected to be the instructor. This sand table terrain model could emit smoke puffs in accord with actual fire commands. This training was to pay dividends within one year.

Southampton Staging Area and Utah Beachhead

A few weeks later we drove to the camouflaged Southampton staging area. The destruction in Southampton was the dawning realization that war was close at hand. We were assigned to the Mahoney-Pitney Liberty Ship for the channel crossing. The United States produced 2,710 Liberty Ships during the war. They were mass produced to serve a short-time need. Many shipyards were involved. Night-time loading of the Liberty Ship, without any lights, and the issue of live ammunition, were accomplished in a few hours. This was one of the few nights when German aircraft did not bomb Southampton Port. The daytime channel crossing concluded with arrival at Utah Beach in Normandy, France. German aircraft met us but their bombs hit farther down the beachhead. Numerous barrage balloons, tethered to the ground with steel cables, floated high in the sky in an attempt to deter German aircraft strafing. We descended rope ladders into waiting landing craft and, when the next tide was right, we disembarked in France.

American forces in Europe peaked at three million. Only 870,000 were in units engaged in fighting such as divisions, artillery battalions, and mechanized cavalry squadrons. Many of of these 870,000 soldiers were in rear echelons of divisions such as ordnance, quartermaster, signal corps, and various headquarters. About 420,000 Americans actually engaged in battle, and 28,000 of these were officers.

Thousands of American soldiers experienced more severe hazards than I. For example, I was not on the D-day landings. Many American soldiers evidenced greater aggressive bravery than I. But there were very few who attained a greater collection of exploits than were destined for those of us in the 283rd Field Artillery Battalion.

This case history, of one of the most unique battalions in the U.S. Army, should have been told years ago. It wasn't. Here we go.

The battalion organization

The 283rd Light Field Artillery Battalion, with about 487 men and officers, consisted of the headquarters battery, medical detachment, service battery, and three firing batteries called "A", "B", and "C". Two revisions of the Table of Organization and Equipment (TO&E) caused some variation from 471 men and officers to 512 in June 1944 and to 482 in June 1945.

Each firing battery had four 105mm howitzers towed by 2 1/2 ton GMC trucks. The cabs of many of the trucks were topped with ring-mounted .50 caliber machine guns. This allowed the gunner 360 degrees field of fire. Firing batteries included sections for radio, telephone wire, vehicular maintenance, kitchen, ammunition supply, and company headquarters. The headquarters section included two forward observer teams of one lieutenant, and a jeep driver. They drew a radio operator from the radio section. The quarter-ton (payload) jeeps usually towed a quarter ton trailer.

Headquarters Battery included a radio section, a telephone-wire section, an automotive maintenance section, a survey section, the all important prestigious fire direction center, the personnel section, the kitchen section, the battery headquarters, individuals to support the staff sections, and the aerial observation section. Our two pilots, Louis P. Nedeau and Frank A. Druyor, assisted by observers Lieutenant Fred B. Whitt and Sergeant Irvin J. Zielenski, used our light aircraft to observe and adjust artillery fire on the Germans.

Captain Frank Cohen, M.D., ran a superb medical detachment of eleven highly dedicated men. Uually, one medic was assigned to work directly with each of the firing batteries.

Service Battery included the 36-man ammunition supply section along with their 2 1/2 ton ammunition trucks usually referred to as the ammunition train. There also were the battalion supply, vehicular maintenance, kitchen sections, and the battery headquarters.

Retention of the primary group

Our soldiers, whose average age was now 20 1/2, were a highly motivated, well disciplined, expertly trained organization. They had a superb sense of purpose and were anxious to get on with their combat role. I don't believe that one of them would have accepted a ticket home. In the instance of the 283rd Field Artillery Battalion there was a central tendency in the distribution of individual attitudes and this became the battalion attitude. We had group solidarity. We not only had very little variance in the men's height but we also had little variance in individual attitudes. This was possibly a fortuitous development, but it may have been linked to some other factor such as the Pennsylvanian orientation. The men knew the United States had intevened in the European war, primarily for our own long-term national security. However, there was far more involved in the formulation of our sense of purpose. Our Jewish soldiers certainly realized the basic evilness of the enemy. Possibly they influenced the other ethnic groups even though these ethnic groups had disintegrated by this time. As we drove deeper into Europe, and observed the atrocious behavior of the German forces, the word "retribution" became more firmly entrenched in our sense of purpose.

We were able to retain our primary group throughout the war. So were numerous other American units. The Germans, on the other hand, were unable to retain the primary groups, which initially were geographically oriented, because of their high casualties on the eastern front (Bartov 1991; p. 18). The strength of the primary group of our battalion may have been stronger than others because of the non-participation in RTC basic training. We had been together as a unit since the men entered the Army.

Allied organization in early August 1944

The two armies which had landed in Normandy on D-Day were the American 1st Army and the British 2nd Army. General Montgomery (British) commanded the 21st Army Group which included both of these armies. Thus, Montgomery commanded all Allied forces in Normandy up until well past the breakout. The Allied forces in Normandy included troops from Poland, France, Belgium, Norway, Czechoslovakia, Denmark, Luxembourg, and the Netherlands as well as the United States, Great Britain, and Canada.

General Eisenhower, the Supreme Commander of all Allied Forces, operated from England until the 1st of September 1944. Eisenhower refrained from telling Montgomery what to do, although he offered suggestions at times. In the meanwhile, the American 3rd Army and the Canadian Army had been assembling in Normandy.

The 1st of August, the American 12th Army Group became operational under General Bradley. It included the American 1st and 3rd Armies. The British 21st Army Group, under Montgomery, now included the British 2nd and the Canadian 1st Armies. Montgomery remained in overall command of both army groups. He also commanded the 21st Army Group. This was the situation when we landed at Utah Beach in Normandy.

Selected for Theatre Artillery

Upon arrival in Normandy, our Battalion was selected to be Theatre Artillery. Normally the heaviest artillery units (those with the largest caliber of guns) are selected for this assignment. This was most unusual for a light artillery battalion. There was need for one rapid-moving artillery battalion which could quickly move to wherever the need was greatest. The honor was based on Army Ground Forces' records of the battalion's performance and attitude in training, along with the battalion fortuitously having been in the right place at the right time. We were the first of eleven non-divisional towed light-artillery battalions to land in Normandy in the month of August (Stanton 1984). We didn't realize it at the time, but in less than a year we would have served with all three army groups that were in the European Theatre of Operations (ETO); all five U. S. armies engaged in combat in the ETO (the U.S. 15th Army not counted because they generally were administering rear echelon troops and reserve divisions); seven of the eleven corps that were engaged in combat before 1 October 1944; eight of the 23 U.S. infantry divisions which were in France before 1 October 1944; and one cavalry squadron. Few, if any, other battalions had so diverse a role in the European Theatre. There was one case where we were assigned to a force screening the left flank of American troops. In all other cases we were assigned to reinforce an offensive operation. In effect, we participated in the majority of major battles, and in the capturing of most major German cities which were within the American zone.

Our itinerary, in the appendix, notes each of the twelve major changes in assignment where we closed down one set of gun positions, moved to a new area, and immediately established new gun positions in support of a different Army or Army Group. At times we wondered who it was at SHAEF (Supreme Headquarters Allied Expeditionary Force) who did such a superb job at keeping our battalion continuously engaged in combat for 268 days (McDonald 1945). Our Battalion was selected to act as an artillery group headquarters in the Rhine River crossing at Worms. Seven other artillery battalions were attached to us and twenty-five additional battalions were on call (Sapp 1945).

The terrain we were to fight over included battlefields dating back to at least medieval times. Gustavus Adolphus (King of Sweden), John Churchill

(Duke of Marlbough), Frederick (Prussian king), Arthur Wellesley (Duke of Wellington), and numerous other military leaders had repeatedly bathed this soil in blood.

Corps artillery, groups, and battalions

The various corps headquarters were occasionally re-assigned to a different army. Generally the armies retained the same corps. The various divisions were reassigned to different corps as the situation changed. Theoretically, an army had three corps and a corps had three divisions. There were five armies and fourteen corps, and so theory regards the number of corps in an army wasn't neglected. There were, at War's end, sixty-one divisions but 15 didn't arrive in France until 1945. Thus, a typical corps had four or five divisions until the later stages of the war.

Divisons had their own organic artillery. Non-divisional battalions of artillery were either assigned as theatre artillery, or else to a corps, possibly through an artillery group headquarters. Battalions were self-sufficient and none were assigned to a group organically. Theoretically there were four battalions for each group. The non-divisional battalions dealt directly with Army on administrative matters and obtained supplies from Army supply points. At one time there were plans for artillery brigade headquarters to command several groups. However, there were few, if any, in the European Theatre of Operations because groups were usually attached directly to a corps.

There was no army nor army-group artillery. Each of the eventual 14 American corps in the European Theatre of Operations had an artillery headquarters unit of 116 men and officers. Also, there were a total of 62 artillery group headquarters batteries eventually in the Theatre (Appendix C lists the numbers of each group, battalion and other details). Thus on the average, four or five artillery groups were assigned to a corps. Each group headquarters unit had 99 men and officers.

There were a total of 260 non-divisional artillery battalions which eventually were in the Theatre. About 21 artillery battalions were selected to be theatre artillery. The other non-divisional battalions were assigned to the various groups, or even the corps artillery headquarters directly. There was flexibility, and when the situation required, or when a corps went into reserve, these assignments were modified. Most of the corps artillery was medium or heavy artillery.

The count of non-divisional artillery battalions at War's end (including both theatre and corps artillery) is in Table 1.

Table 1.
Non-Divisional Artillery Battalions in the ETO

75mm Parachute	3 battalions
Armored 105 mm	16 battalions
Towed 105 mm howitzers	37 battalions
155 mm howitzers	72 battalions
4.5-inch gun	17 battalions
155mm gun	38 battalions
8-inch howitzers	37 battalions
8-inch gun	5 battalions
240 mm howitzers	15 battalions
Captured enemy wpns.	1 battalions
Field Arty Observation	19 battalions
Total non-divisional battalions	260

Information compiled from and reprinted with permission from *Order of Battle World War II* by Shelby L. Stanton c1984
Published by Presidio Press, 505 B San Marin Drive, Novato, CA 94945

Divisions and their artillery

The more fully the infantry division adhered to McNair's "sound fundamental" and carried as part of its organization strength only the men and equipment it would need under practically all conditions, the lighter and more mobile the division became (Weigley 1981; p. 23).

At the end of June 1944, there were 13 American divisions in France. In March 1945, there were 61 American divisions in the Theatre (Stanton 1984). Each division had its own artillery battalions.

Table 2:
Arrival of U.S. Divisions in France

	Infantry	Armored	Airborne	Cumulative Total
June	1 2 4 9 29 30 79 83 90	2 3	82 101	13
July	5 8 28 35	4 5 6		20
Aug.	3* 36* 45* 80	7		25
Sept.	26 44 94 95 102 104			31
Oct.	100 103	9 14		35
Nov.	78 84 99	12		39
Dec.	66 75 87 106	10 11 17		46
Jan.	42 63 65 69 70 76 89	8 13		55
Feb.	71	16 20	13	59
Mar.	86 97			61

* landed in southern France

Information compiled from and reprinted with permission from *Order of Battle World War II* by Shelby L. Stanton c1984 Published by Presidio Press, 505 B San Marin Drive, Novato, CA 94945

An infantry division had three artillery battalions, each with 12 towed 105mm howitzers, plus one battalion of 12 towed 155mm howitzers. An armored division had three battalions, each with 18 mechanized 105mm howitzers. An airborne division had three battalions, each with twelve 75mm pack howitzers. Thus, by March 1945 there were 224 artillery battalions which were organic to the divisions. All of this was light artillery except for the 155mm howitzer battalions. They were medium artillery.

Total artillery in the theater by war's end

The 260 non-divisional battalions plus the 224 divisional battalions total 484. However, the divisional battalions were mostly light artillery whereas the non-divisional were mostly medium and heavy artillery. The battalion numbers of the non-divisional battalions are in Appendix C. There were about 263,658 artillery men and officers in the Theatre by War's end (see Appendix C). When casualties and replacements are considered, I would estimate there were about 282,000 men and officers. This is 9.4 percent of the total of American forces. Yet, they inflicted far more than half of the total German casualties. Some writers (Deighton 1993:p. 483) suggest that 90% of German casualties on the western front were from artillery. Artillery, and particularly light artillery, could be used in close proximity to friendly troops.

Eradicating Germans from hedgerows

We were assigned the radio code word of Highhedge. The battalion commander was Highhedge 6. The commander of C battery was Highhedge Charlie 6. Most every officer had a Highhedge code name/number. This particular code name reflected the preponderance of high hedges which prevailed along boundaries of most every field in Normandy. In some cases there were two closely parallel hedge rows with a narrow sunken farm road in between. Hedge rows were a result of hundreds of years of farming the same fields, and never plowing the edges of the fields. In effect, the hedges had been protected from erosion, and in addition they had accumulated centuries of decomposed leaf litter and other debris. Each one consisted of about six feet of soil topped by tree and plant growth. German soldiers had dug their shelters, known as foxholes, into these banks knowing they would not be inundated by heavy rains, and that they would have considerable protection from bomb and artillery fragments. One reason the Normandy campaign dragged out for eight weeks was the difficulty of eradicating German troops from the hedgerows. Although the slow movement through the hedgerow country was causing many more American casualties than had been anticipated, German losses were far greater. The extent of the German losses were unknown to the Allies at the time.

The overwhelming numbers of combat casualties since the Civil War have been from artillery fragments (shrapnel). It seemed that about half the killed in action were from head wounds. Steel helmets were some protection as was attested to by dented helmets. However, artillery and bomb fragments can, and did, penetrate numerous helmets. Despite television portrayals of combat, few soldiers are killed by bullets, and even fewer are killed by bayonets or knives.

Vehicular markings

Unit markings on vehicle bumpers were blacked out for security reasons. However, our battalion instigated a logo of a horizontal rectangle with three blocks colored yellow, red, yellow. The logo was painted on front and rear of every vehicle, howitzer, and trailer. The logo was also used as a directional sign into command posts. Few, if any, other units employed a similar method of rapid identification. It was extremely useful for ready recognition in poor visibility conditions. It facilitated our long distance moves by identifying our vehicles as part of a rapid moving high-priority unit for the military police directing traffic. I found it an invaluable technique in locating my jeep amongst many others when I returned to the vehicular convoy after having been on foot as a forward observer. Divisional units wore their division patch on their left shoulders. We did not wear any shoulder patch.

The short spell in Third Army

Our battalion was assigned to Third Army upon landing at Utah Beach. Third Army, commanded by General George Patton, had not been in the Normandy invasion. Now, they were assembling in Normandy for the purpose of spearheading the breakout from Normandy. The key to the breakout of Normandy was Avranches at the base of the Peninsula. It was also the entrance into the Brittany Peninsula. Preinvasion plans had specified that the capture of Avranches would trigger the Third Army's entrance into the battle. Avranches was captured by the American Fourth and Sixth Armored Divisions on July 30th. From Utah Beach our battalion went to the XII Corps marshalling area where the sounds of artillery fire were clearly audible. It was impressive to realize the reality of the cannon firing. Twelfth Corps was a part of the Third Army.

We were assigned to General Patton's Third Army only for those first two days. General Patton allowed reporters greater latitude than the other army commanders. For example, reporters could mention the name of towns and cities as soon as Patton's troops arrived, whereas other armies required a 48-hour delay. Reporters quickly learned they could cite towns and cities immediately upon capture, providing they gave credit to Third Army. They were inclined to do this when army boundaries were close, and they could profess to have confused the boundary. This was particulary evident to us when we captured Nuremberg and Third Army received the credit in our hometown newspapers.

Mortain, First Army, 30th Division

The capture of Avranches on July 30th was most disturbing to Hitler. It meant that the Allies had broken out of Normandy and if a war of rapid movement followed, the German Army would be at a great disadvantage with their horse-drawn wagons and artillery. Hitler seized upon this loss to demonstrate his military tactical ability. By his plan, German troops would divide the Allied Forces with a strong attack from Mortain to Avranches. Patton's Third Army would be cut off and then destroyed. Hitler then planned to drive the allies back into the Ocean.

Mortain had been captured by the First Division on 3 August. The 30th Division replaced them in their foxholes and other defenses on the 6th of August.

A few hours later, in the early morning hours of 7 August, the German attack began. However, the critical terrain of Hill 317 which dominated Mortain was defended by troops who were to demolish Hitler's plan. One of the heroes of the episode happened to be Robert Warnick, a high school classmate. He was featured in numerous photographs, including in Blumenson's book.

> The fact that the 2nd Battalion, 120th Infantry had retained possession of the top of Hill 317 was one of the outstanding small unit achievements in the course of the campaign in western Europe. The surrounded force for five days denied the Germans terrain that would have given them observation of the VII Corps sector (Blumenson 1963:p. 246).

German forces withdrew to recover from the severe mauling they had suffered at the hands of American artillery, much of which had been directed from Hill 317. Ostensibly, they were to be reinforced with additional Panzer divisions and then try again on the 20th of August. They never did.

On August 11th, in deference to recommendations from Bradley, and also in recognition of the possibility of entrapping huge German forces in the Falaise-Argentan area, Montgomery issued a directive with plans for the proposed entrapment of German forces. The 30th Division was in an excellent location to participate in the new operation. They were scheduled to move from Mortain to Domfront, where they would become the anchor force at the most western end of the planned pocket. They could well use the 283rd Field Artillery Battalion because they would be in a position to shell the large German forces within the proposed pocket. Our orders assigning us to Third Army were rescinded. Now, we were in direct support of the 30th Division, XIX Corps, and First Army. This sudden change in assignment was to be the pattern for the entire war. Our battalion staff reacted to these changing assignments with amazing rapidity.

The 30th Infantry Division

A full-strength infantry division had 14,253 men, warrant officers, and officers. The 30th was originally a National Guard division from the Carolinas, Georgia, and Tennessee. When they landed at Omaha Beach, on the 10th of June, they had taken in thousands of filler personnel from many states. The Division suffered many casualties in the hedgerow fighting. By the 1st of August, they had picked up more than a thousand replacements. At one time or another the 283rd Field Artillery Battalion was in support of each of the infantry regiments (117th, 119th, and 120th). Major General Leland S. Hobbs had commanded the Division since 1942. The 30th Infantry Division was often teamed up with the 2nd Armored Division. The Division was assigned to First Army until Ninth Army became active on the western front in October. They were assigned to many corps.

The 30th Infantry Division was rated first in overall combat performance by ETO military historians according to the following letter from Colonel S.L.A. Marshall to General Hobbs. The aforementioned Captain Robert Warnick was given a copy of the letter by General Hobbs and has held it for nearly fifty years. Knowing Warnick as well as I do I truly believe it is authentic. To my knowledge it has never been published before.

16 March 1946

Dear General Hobbs:

Now that I am leaving the service, I thought it might be well to give you the following information for whatever satisfaction you may derive therefrom.

I was historian of the ETO. Toward the end of last fall, for the purpose of breaking the log-jam of paper concerning division presidential unit citations, General Eisenhower instructed me to draw up a rating sheet on the divisions. This entailed in the actual processing that we had to go over the total work of all the more experienced divisions, infantry and armor, and report back to him which divisions we considered had performed the most consistent battle service.

In our report we named certain infantry divisions in the first category and the same with armor, and we placed others in a second category and yet others in a third. The 30th was among five divisions in the first category.

However, we placed the 30th Division no. 1 on the list of first divisions. It was the combined judgment of the approximately 35 historical officers who had worked on the record; and in the field that the 30th had merited this distinction. It was our finding that the 30th had been outstanding in three operations and that we could consistently recommend it for citation on any one of these three occasions. It was found further that it had in no single instance performed discreditably or weakly when considered against the averages of the theatre and that in no single operation had it carried less than its share of the burden or looked bad when compared with the forces on its flanks. We were especially impressed with the fact that it had consistently achieved results without undue wasteage of its men.

I do not know whether any further honors will come to the 30th, I hope that they do, for we had to keep looking at the balance of things always and we felt that the 30th was the outstanding infantry division of the ETO.

Respectfully yours,

s/S. L. A. Marshall

Colonel S. L. A. Marshall, GSC
Historian of the ETO

Millny, Normandy

Our first Battery "C" gun position was near Millny on the Normandy Peninsula. In deference to the few German aircraft which ventured into our skies, the howitzers and other battery installations were camouflaged under nets, interwoven with green and brown burlap strips. In order to improve on the camouflage of our first position, the men added freshly mown hay from the fields. A French farmer appeared shouting and gesticulating with his hands. It was obvious that he was concerned about the loss of his hay. First Sergeant Charles Mitchell reached into his pocket and withdrew a printed T.S. card which was good for ten free trips to the Chaplain. He marked on the card that it was good for three loads of hay and signed his name. The farmer immediately adjusted his attitude, accepted the card, stated many "Mercis", raced to his house and came back with a bottle of Calvados, a strong apple brandy. Obviously he accepted the card as an official military requisition for which he would eventually be paid.

Probably, when the farmer eventually presented the card to allied military government, he did get paid after a hearty laugh. The origin of T.S. cards is unknown, but during the war era when soldiers complained about most anything someone would suggest he tell it to the Chaplain. Others would simply say "tough s — -". The two were combined and most first sergeants had a supply of printed T.S. cards.

At this stage of the war all aircraft were propeller driven. German aircraft usually ventured into our skies only during darkness. Their noisy engines, similar to a model 1920 washing machine, announced arrival.

Ammunition procurement and an accidental death

We were in the hedgerows of Normandy when Captain Sapp told me to unload the ammunition train trucks and go back to the Army ammunition dump and load up with as much as we could carry. He said the Germans were expected to counter-attack. I told the Lieutenant at the dump that we were there to load up. He replied that without signed paper work we couldn't get any ammunition. No, contacting Captain Sapp on the radio wouldn't suffice. We needed paper. Spallone and I begged and negotiated to no avail. Eventually, Sergeant Patrick came in the tent and said we had better leave. Outside the tent, he told me that he and the fellows had stolen the ammunition. Our trucks were full (Parker 1994).

It was a dark night when we arrived at the entrance of our hedgerow. I got out of the Command Car and walked in front of it with the trucks following and using only black-out lights. I walked along the inside of the perimeter. There was no indication that anyone was there. A shot was fired and I thought we had walked into a trap. Everyone jumped out of their trucks and no one knew what was happening. There was dead silence. Then we realized that Sergeant King, who had been with us, had been shot when he had left his truck and walked across a field. He was killed by one of our own guards (Parker 1994).

(Ed. comment) The maturity and calmness of Sergeant Parker and his associates avoided a more deadly fire fight in the dark Normandy night. Fortunately, they did not assume the shot had been from Germans. Had they done so, it could have led to many casualties.

The number of American soldiers killed accidentally, including by misoriented American airplanes, and short artillery rounds from American howitzers, was large. Aircraft accidents, including mis-oriented bombing, accounted for 21,821 deaths (Army wide). Other accidents caused 22,014 deaths (Beebe 1952). Fortunately, the 283rd Field Artillery Battalion was never involved in American casualties from short artillery shells. This is truly remarkable when one considers the more than 45,000 rounds which were fired, and each shell was propelled by as many as seven powder charges. Light artillery range calculations are a function of both the angle of the artillery tube, and the number of powder charges used. Each round comes with seven clearly labeled, bagged, powder charges. If the range calculation is charge 5, then the gunners remove both charge 6 and charge 7, hold them in the air until verified, and then fire with the shell being propelled by the remaining five bags. Unused charges are piled and eventually burned. Fire missions were received at most anytime, including the early hours of the morning when only two or three gunners might be on duty at each gun position. Our Pennsylvania gunners had the intelligence and determination to make no errors at all. One error out of more than 45,000 rounds fired would have been one too many. Artillery gunners must be better than ordinary people. Our Army's intelligence standards are essential.

Chaplains

Our troops usually had access to a Chaplain. Combined Protestant-Catholic services were held when possible. There was usually a Jewish chaplain nearby for our Jewish soldiers. Again, the relationships between soldiers of various religions was superb. Chaplains were likely to be involved in ministering to the dead and wounded, as well as corresponding with grieving families.

The Falaise-Argentan Pocket

My first combat assignment, like all the others, was extremely brief. This allowed me all the latitude I could have wished for. Overall, my Army experience has been that as long as I was doing well, I had more freedom of action than foresters have. In this one case, on the 15th of August, I was told to report to the Command Post of the Reconnaissance Squadron of the 30th Division, and to support them by adjusting required artillery fire. Domfront was under German artillery fire, so I parked my jeep at the edge of town, where I felt my two men would be safe, while I walked on in. I was wrong, and Jacques Linn and Leonard Janssen had to relocate when they came under German fire.

Although many walls remained standing, Domfront was in nearly complete destruction. The rubble from the destroyed buildings concealed the location of former streets. Most of the houses had been farm homes. In peace times, the farmers went forth to their fields on a daily basis. Animal barns were attached to the homes. The manure was usually piled in the front yard. Now it was all mixed with the rubble. Farming in France was not at all like farming in America.

The Domfront command post was in the ruins of a farmhouse. The aid station was in the basement. Litter bearers were bringing in seriously wounded. It was not a pretty sight for a new arrival in combat. The squadron commander, a lieutenant colonel, was nonchalantly shaving. He welcomed me and suggested

I locate some of the German guns and silence them. He provided minimal additional information.

German snipers were trained to search for officers as targets. Thus I did not wear my lieutenant's insignia. I carried a rifle or carbine as well as my pistol and a radio. Radios were problematic. The small hand held radios had a range of less than two miles. The larger radio, with its battery pack, was heavy and cumbersome . The jeep-mounted radio could usually reach battery and battalion headquarters. In order to communicate with my jeep, where the radio operator would transmit my fire commands, I chose to carry the radio which seemed best suited to the situation. Neither one was very good.

In my dealing with strange troops it was odd that I was never challenged in any way. Requested information was readily granted. Probably personnel turnover was so rapid in infantry units that few men knew many of their own officers. At Domfront, I was guided up towards one front-line company, but no one knew where the company commander was. With assistance from a corporal, I observed a German artillery position. I attempted to adjust artillery fire on them, but I had difficulty making radio contact with my crew because, as it turned out, they were under German fire and were relocating our jeep to a safer location. Fortunately, I made contact with our fire-direction center. I reported the coordinates from our 1:25,000 black and white maps and then adjusted from there. We neutralized the German position. Shells, shell casings, and howitzer parts were blown high into the sky. Most of the gunners were killed. That was only the first of a series of missions.

> The situation (for the Germans) worsened that day (15 August).
> Allied attacks continued, with Falaise, Domfront, and Argentan the critical points of pressure (Blumenson 1963:p.269).

The Falaise-Argentan pocket became a cauldron of devastation as Allied Forces narrowed the escape hatch between Falaise and Argentan. On 15 August the pocket was about forty miles long and fifteen miles wide (Blumenson 1963;p. 277). Allied forces slowly narrowed the gap, and surrounded the pocket with artillery which proceeded to cause havoc amongst the Germans. Most of our artillery fire was observed fire from our forward observers and especially our aerial spotter plane. During those five days there was usually ground fog in early mornings which enabled German forces to move undetected. However, the ground fog tended to lift rather suddenly exposing huge German forces moving easterly. Then they became as good a target as sitting ducks and we had no qualms about shooting sitting ducks in this case. The escape hatch was closed on 17 August although a few German forces continued to infiltrate out of the pocket.

Some texts have focused on the tactical errors which caused a delay in the closing of the gap and allowed some German units to escape. There were numerous tactical and logistical errors during the war. They caused unnecessary American casualties. However, if the author begins citing each case we might well miss the objective of this narrative. The sum of these errors are clear evidence of the need for soldiers and officers with high intelligence in our Armed Forces.

Eventually, Allied Forces had to work through the pocket accepting surrender of the entrapped Germans in large units, small units, and even individuals. By August 19 the pocket was a devastating defeat for German forces. Ever onward, we moved through the pocket on August 20th travelling 109 miles in one day.

The Artillery Battalion Commander supervises his observers

Lieutenant Colonel McDonald, our battalion commander, was the antithesis of "Dugout Willie". In this first battle in our European action, he set out to locate his forward observers. He found one from Battery "A" and provided some guidance. Then Colonel McDonald, his radio operator, and his C&R car driver started their return to the battalion command post but became confused with the road network where all road signs had been removed by German forces. Nightfall was approaching. The radio operator well recollects the event.

> The driver stopped at a dirt road. I was told to go down the road on foot to investigate. I asked for the driver to come along. Request denied. I was given a direct order to find out what was down that road. So, a scared young soldier cautiously obeyed and started down the road. I went about 100 yards and looked around for a rock. I threw it as far down that road as I could, straight into the pitch black of night.

> As the rock fell with a thud, a machine gun gave a burst of red and white. My reflexes spun me around and over to the side of the road. I found myself in a gully. I lobbed off two grenades and hightailed it back to the command car. Needless to say we didn't take that road (Pensock 1993).

54

Shattering myths of German mechanization and military excellence

Forty thousand Germans were captured and 30,000 were killed. Burned and wrecked German tanks, vehicles, equipment, and army horses littered thousands of acres. The myth of a modern German Army was shattered. It seemed as though most of their forces were still using horse-drawn wagons and horse-drawn artillery. The stench of the decaying human and horse flesh, aggravated by August heat, could even be smelled in overhead aircraft. Long lines of German prisoners, marching four abreast, extended along the roads as far as eye could see. Some were dazed, but others seemed to be happy to be out of the war. Perhaps they were aware of their forthcoming life-time opportunity of a free round trip to the United States with good meals and good accomodations. Thereafter, the sight of dead soldiers was commonplace. Their burial was no responsibility of combat troops. Graves registration units would eventually be along. The elite German 2nd Panzer Division attributed their defeat to the accuracy and the speed of American artillery. Artillery promptly impressed friend and foe alike as the outstanding branch of the American armed forces (Weigley, 1981:p.127).

> In combat the 105mm howitzer fully justified all the study, work, and hope which had been placed in it. From Alaska came an early report, "the 105mm field howitzer, despite almost 'round-the-clock firing in wet weather with no time for usual maintenance except an occasional bore swab, functioned without difficulty and maintained its long-range accuracy". German troops were so impressed by the rate of American fire that prisoners often asked to see the "belt-fed" or "automatic" artillery. American artillerymen proudly referred to it as the "back-bone of the American Artillery." (Sawicki 1978: p. 1240).

Caught inside the Falaise Argentan pocket were
  Five SS panzer divisions
  Six Army panzer divisions
  Eight infantry divisions

(Stein, 1966;p. 225)

Our great drive into northern France

Domfront was the launching pad for our great 16-day-drive through northern France into Belgium. Our battery was somewhat in advance of our battalion headquarters, and I, being a forward observer with the infantry, was well in advance of our battery. Numerous gun positions were occupied each day of this drive, usually firing artillery at rear-guard Germans and snipers. There were gun positions where no rounds were fired. It is difficult to correlate our travels with modern maps because some towns were virtually wiped off the map and have never re-appeared. Lieutenant Fred Whitt, of "B" Battery kept a listing of 168 towns we passed through, but not including dates. Even this list does not include the many towns we passed through on our distance travel on the way to the Battle of Colmar. The author can supply the "Fred Whitt" list on request. The itinerary in the appendix is from the official morning reports and reflects the position of the battalion headquarters and "C" Battery at the start of each new day. It does not name all of the towns and cities we passed through, nor all of our gun positions. Morning-report positions were usually small towns, whereas the towns/cities we passed through were often large cities such as Evereaux, Chambly, Cambrai, and Vallenciennes. I will strive to include the larger cities in the narrative. Our route is indicated on Map 5.

Battery C of the 283rd Field Artillery Battalion departed Domfront 20 August, travelled 109 miles to Crucey.

21 August, travelled 18 miles to LeSouchet

23 August, travelled 12 miles to Evereaux

25 August, travelled 13 miles to Verdun
(Not the well-known Verdun of World War I.)

27 August, travelled 25 miles to Mantes-Gassicourt and crossed the Seine River on a pontoon bridge.
Then passed through Limays and Chambly.

30 August, travelled 28 miles to Gennicourt

31 August, travelled 18 miles to Neuilly
Then passed through Belle, LeTillet, Liamcourt, Cambrai, and Valenciennes.

2 September, travelled 125 miles to St. Maur, Belgium
Total 348 miles or 22 miles per day for 16 days
Between Domfront and St. Maur we passed through 78 towns/cities for an average of one town/city every four and a half miles and five towns/cities per day (Whitt 1945).

The French homefront

No one in "C" battery spoke French or German. Few of the rural French people we met spoke any English. One French woman, who spoke some English, related how her soldier husband had been taken prisoner by the Germans in 1940 and since then had been required to work in a German munitions factory. The number of French soldiers captured in 1940, and transferred to Germany was about 1,580,000. A few were eventually released in exchange for young volunteer Frenchmen. Many, including her husband, had been granted a pass to visit their homes for the 1942 Christmas season. However, so many never returned to their work in Germany that no further passes were granted. The French homefront was not a happy place. More than one million fatherless families became the domain of the mother, who reluctantly assumed the family leadership role and the financial responsibility. This woman's story made a deep impression on me because I had believed slavery, in this world, had ended. These were not prisoners of war because they were required to work in munitions factories and other war effort. Furthermore, the French and the Germans were no longer at war.

Deighton reports the French 1940 casualties as 1,900,000 prisoners and missing; 200,000 wounded; and 90,000 killed (Deighton 1993: Table 3). It is difficult to imagine the military ineptitude which permitted so many prisoners to be taken. The French defeat might partially be attributed to their heavy losses in World War I. When 27 percent of all a nation's men of ages 18 through 26 are killed (Willmott 1989:p.87) the ensuing generation may well lack passionate leaders.

The enslaved French men were compelled to work hard. They had no choice. Undercover Gestapo agents investigated prisoner-of-war malingering. In the case of the coal mine at Concordia several victims were singled out and sent to Dachau concentration camp (Gillingham 1985).

Germany had also enslaved millions of Poles, Czechs, and Russians. There was no need for German women to engage in war work until the closing days of the war. They were free for household work, and in cases even had household help. Some of the enslaved persons, later known as displaced persons (DP's), chose to remain in Germany after the war.

Point men

I walked much of the distance after Domfront, accompanying our Infantry forces. Decisions, as to when and how to move forward and where to place my jeep in the column, were usually for me to make. I had remarkable freedom of action. Generally, in the advance I would accompany the three-man point.

Later, when we were with the 29th Division, that was required. Usually a fine young corporal led the way, obtaining his instruction by radio. I was usually able to veer the point off the road and into the woods, walking parallel to the road. It seemed far safer to me.

There were times when the point and I were pinned to the ground by German fire and had to lie low until we were rescued by other artillery observers or concentrated American fire. The higher rate of fire of German machine guns emitted a unique sound easily differentiated from American guns. German troops were usually more anxious to fire at the main force, which they knew would be following the point. The situation was tantamount to a land survey crew in the Appalachians. The lead chainman stirs up the yellow jackets and copper head snakes so that they are fully ready for the main survey crew. I am not going to delve into the grisly scenes which were our every day fare. There were numerous fine Americans killed and seriously wounded. It was not a pretty picture and certainly not glamorous.

A typical day on the offensive

A typical day on this attack started at 4 a.m. when breakfast was served by the Infantry from a kitchen truck. There were mugs of hot coffee whitened by canned Carnation milk if that is what you wanted. The manner in which the powdered eggs were cooked denoted the quality of kitchen troops. They were easily baked leaving little taste. They could also be whipped into a batter and fried like hot cakes, which were tasty. The Army had devised a method of baking bread, somewheres in the rear echelons, which also preserved it for a couple of weeks. Tasteless, until topped with canned butter and jam. My appetite was voracious. I always packed a few "D" bars which were potent high-energy chocolate, readily eaten while on the move. I disliked the canned "C" rations (hash, stew etc.) but nevertheless ate many cold, and washed down with water from my canteen. My teams were sometimes able to scrounge boxes of 10 in 1 rations which were far better. There were crackers, canned cheese, canned Spam, fruit bars, etc. Some days, at lunch time, I dropped back to my Jeep and had a good meal.

By morning twilight the infantrymen had rolled and stacked their sleeping bags. These would be hauled by vehicle. My bedroll would be in our jeep trailer. The word would come to move out and each infantryman seemed to know what to do. I would also move out on foot, knowing that my jeep driver and radio operator would be bringing the jeep along at an appropriate place in the column. Most important, we would verify our radio communications before moving out.

One of the better items of issue equipment was my heavy canvas bed roll. I still have it. During the war, I overlapped three wool blankets inside of it and secured them with large blanket safety pins. There are pockets in the bed roll for dry socks and other clothing. Many of us received dry socks from home. I never used a mattress, but simply unrolled the bed roll on the ground and could then sleep most anywhere when time permitted. Frequently, I would douse the inside with powdered DDT in order to prevent lice from becoming problematic. The powder, in a small gray can which came along with the rations, smelled good and killed lice and most every other insect. I wondered why we had not had it for those Louisiana chiggers. In World War I, typhus and other lice-carried diseases had caused many American casualties. None of these diseases plagued our World War II forces, thanks to DDT. Army-wide wartime deaths from disease were 14,243, which although, too large, was smaller than the 51,447 in World War I, and far smaller than the 233,789 Yankee soldiers who died of disease in the Civil War.

French villages, towns, and cities

Most German resistance was in rural wooded areas. French Resistance men were taking up their hidden rifles and greeting the Americans. They also reported known German positions which virtually assured us that towns and cities were clear. There were 78 villages, towns, and cities in France, and more in Belgium and Holland, on our agenda (Witt 1945). When we approached towns and cities, I would drop back to my jeep, and then ride in style into town amidst the shreiks of "Jeep" "Jeep." Yes, I was often the first American vehicle to arrive. I never did find out how all these people knew the word "jeep".

Virtually every inhabitant turned out to welcome their liberators in what amounted to a several-days-long ovation. Hundreds of boys and girls, working shifts on the ropes, set all the church bells to ringing for several days, with never a missed beat. There were tear-streaked old men who had never thought they would once again be free. Flower-throwing old women were elated. Many women had endured five lonely years while their husbands were in German captivity. Now they had hope. Skinny kids were simply awed because few of them had any remembrance of the day when they had no fear of every uniform. All the townsfolk were skinny. There were no fat people at all. The Germans had systematically plundered the entire economy of France, Belgium, and, worst of all, Holland. Obviously, the townfolk were malnourished. Babies in their mother's arms, with eyes as big as saucers, realized that something was up. In many cases the event became their first memory. There were few, if any, dogs. For most everyone, and especially our soldiers, liberation day would endure in their memories forever.

Our first howitzers were greeted by a thunderous ovation which resembled an artillery barrage as it rolled from one end of town to the other. Our howitzers were seen as a sure sign that the townfolk had seen the last of war. At the same time, off in the distance the muffled roar of cannon were reminders that other people were still to be liberated. Seldom were our trucks able to stop. Amazingly, the populace made way for our guns to keep rolling because they realized there were many more towns to liberate. Our trucks were bedecked with flowers. The exhilaration, which came from seeing so many people exuberant over their newly-regained freedom which our soldiers gave them, had a most profound influence, and for a life-time duration, on our men. Church bells will forever be a remembrance. Doing something really worthwhile, day in and day out, overcomes the forces of evil. Generations of Pennsylvanians have benefited morally from these experiences.

Our divisions, although still outnumbered by German divisions, were on a "roll". Events were moving so fast that these newly liberated towns were in the Allied Forces rear area within a day or so.

Retribution for German collaborators

There was a dark side to the deliriously joyful scenes in recently liberated parts of France. Immediate retaliation against people who had collaborated with the occupying German forces resulted in some injustice. Roadside trials were held and immediate punishment imposed. We saw numerous cases of women having their heads shaved because they reputedly fraternized with German troops. There were cases where evidence was stretched, or even falsified, in order to arrange the immediate death of some neighbor who had been disagreeable in the past. This was all topped off by political splits within the French partisan forces.

> In the decade following the liberation, the purge was often compared to the reign of terror during the French Revolution (Lottman 1986; p.16).

> Summary wartime executions of accused collaborators by the Resistance gave way after V-E Day to a more systematic and legal purge. French courts sentenced 120,000 collaborators and executed 2,000 (Botting 1983: p.23).

In one case where there was a trial, the court clerk ordered 75 coffins from a local carpenter before the trial began (Lottman 1986, p. 117). The case of Jean Giono, one of the foremost French writers of that period, is illustrative of the happenings. Giono had fought with the French in World War I, and had then become a pacifist. On the outbreak of war in 1939, Giono was imprisoned in Marseilles because of his pacifist activities. Giono's arrest aroused a storm of protest from throughout the world, including students at Yale University. He was released from prison in November. During the war years, Giono remained at home writing books. After the liberation of southern France in 1944, the Communists began a reign of terror in the region. Giono was suspect on two accounts; he had never spoken out against the German occupation nor the Vichy government, and he had published an interview in a collaborator's journal. Giono was arrested and confined six months in the prison at Digne. When the communists were ousted from the region, Giono was freed without trial. (Smith, Maxwell). Giono's writings include *The Man Who Planted Trees*.

LeSouchet

When the unit I was with approached LeSouchet, French partisans warned us of mines. Retreating Germans had laid anti-tank mines on the paved road and covered them with straw. The French helped us in the mine removal, which was not very difficult in this case. This road, like most rural French roads was lined with Lombardy poplar trees.

The thirty-five mile advance

There was one day when we walked 35 miles, occasionally encountering German resistance. German snipers would fire a few rounds, possibly killing an American Infantryman, then throw down their rifles and surrender. Protecting the lives of these prisoners was extremely challenging for our officers. The Infantry battalions were assigned lettered phase lines and a route of advance each day. On this one day, whenever the infantry battalion commander reported they had reached a phase line, the response was to go another phase line, and then another, until near dark. Each order was greeted with moans from the tired infantry. Northern France is at a latitude of 51 degrees or the same as Calgary, Canada. Summer days were longer than any in the lower 48 states.

The miles of walking to school, regardless of the weather, is a classic example of adversity preparing one more G.I. for the ensuing war. The Army's 25-mile forced marches, in training, were easy for me. I used to overcapitalize on this by amazing the troops in going from front to rear of the column, and then back to the front. In northern France, I had no trouble in maintaining pace with the Infantry advance.

The mechanics of walking

I also had a spring in my step that few other people had. Back in Camp

Shelby I had noticed many soldiers would wear out the soles of their boots in two or three months, whereas, mine never wore out. Strangely, it meant that in parades, although I was in step, I was going up as the others were going down. This phenomena did get me out of most parades at Camp Shelby. I have observed other people who walk in this manner although they probably number only one in a thousand. In 1993 I am able to walk many miles without any thought about the matter. The mechanics of walking was researched, using German soldiers, in 1894 (Weber 1894). Much of the research since then, concerning the kinetics of walking, has been published by Springer-Verlag, the foremost German scientific publisher. We'll encounter the wartime editor of Springer-Verlag in the case history under "The Oslo Papers".

"Sergeant Maupin had a similar gait (Davis 1993)."

Reconnoitering gun positions and command posts

Our artillery battalion commander, Lieutenant Colonel Hugh McDonald, was superb at reconnoitering battalion command posts and howitzer positions. He always used his own initiative, striving to advance as far as possible each time so as to minimize the eventual number of moves. By the end of the war he had reconnoitered 76 battalion command posts, and upwards of 300 howitzer positions. He usually rode in his 3/4 ton-capacity command and reconnaissance car (C&R car). This was necessary because a jeep was not able to carry the very large radio he needed. It was an impressive, high-framed convertible vehicle with a substantial trunk. He did not tow a trailer because there were occasions when his driver had to make a quick turnabout. Much of his recon was on foot. He carried an M1 rifle along with ample ammunition, a .22 caliber Luger in a shoulder holster, and an Italian machine gun in his car. There were times when he was in advance of our front-line troops which he shouldn't have done. He seemed to be fearless, yet his survival is evidence that he used judgment.

He was usually followed by the three firing battery commanders in their C&R cars or jeeps. Colonel McDonald would quickly sense the situation, assign general areas for the firing batteries, and then release the battery commanders so they could arrange to issue the close station march order (CSMO) to their executive officers. Then Colonel McDonald proceeded to select an appropriate battalion CP. When there were large assemblies of artillery battalions, to support a river crossing, the need for firing battery positions was larger than the available supply. It came down to whoever was there first had the site. Thus, once assigned a site, battery commanders would not leave until their battery was in the new position. Movement arrangements were by radio. The courteous cooperation between units competing for various sites was superb.

The reconnaissance party, with Colonel McDonald and the battery commanders, ran into a nest of Germans at Gennicourt. The score was home-team: No casualties. Visiting Germans: 3 killed, 12 wounded, and 22 prisoners (McDonald 1945).

When we drew enemy counter battery shelling, it was usually better to dig in deeper, than to try to move elsewhere. Batteries which tried to move were frequently caught with all personnel exposed to enemy fire.

The American Third Army's great encirclement

While the 21st Army Group (British and Canadian forces), the American First Army, and parts of XV Corps of Third Army were engaged in conquering the German forces in the Falaise-Argentan Pocket, the remainder of the American Third Army, under General Patton, drove easterly and then northerly to the Seine River in an attempt to entrap those German forces who might have escaped the Falaise-Argentan Pocket. Patton also used the situation to drive his XX and XII Corps towards Chartres and Orleans, southwest of Paris.

The Seine River crossing

Mantes Gassicourt is on the Seine River, about thirty miles west of Paris. Supposedly, it was just inside the terrain assigned to the British Army. The 79th Division (part of Third Army and XV Corps) arrived there and established a beachhead on the far side of the 700-foot wide river on 19 August, the same day First Army was concluding the Falaise-Argentan pocket. This meant that First Army would have been pinched out of the front line because they had been between the British and Third Army. Thus XV Corps was temporarily assigned to First Army, and the 30th Division was reassigned to the XV Corps. Our 283rd Field Artillery Battalion continued with the 30th Division.

On the 27th of August the 30th Division arrived at Mantes Gassicourt, crossed the Seine River on a treadway bridge, and passed through the 79th Division. Two days later, the 79th Division and XV Corps reverted to Third Army and once again we were with the XIX Corps.

Nervous Nellie

One night, near Mantes Gassicourt, twenty of us from Service Battery bedded down in a chateau after posting guards. We chose to sleep in a long narrow hall with our heads against an interior wall. During the night we were awakend by shots and someone shouting "Halt". Instantly we were all awake, and quite concerned about what might be happening. We found out that one of our guards had mistakenly shot up the iron-clad statues at the end of the hall (Parker 1994).

(Ed. comment) Few people can understand what one's imagination can do to you in a combat situation, even though you are not threatened.

The source of fire missions

Most of our fire missions were unobserved fire where the targets were derived from intelligence, aerial photos, artillery observation battalions, and other sources. Division artillery headquarters would assign these targets to our fire direction center.

Some of our observed fire was from Lieutenant Louis Nedeau, or Lieutenant Frank Druyor in our Piper Cub aircraft, model L-4. Usually Lieutenant Fred Whitt or Sergeant Zeke Zielinski was with him. However, in deference to the highly accurate, and powerful, German 88mm, 40mm, and 20mm anti-aircraft guns, they seldom ventured over the front lines. "Druyor did take a bullet in the rear end" (Davis 1993).

Our six forward observers brought in the most meaningful targets. Overall, artillery forward observering is a dangerous occupation. On the other hand, the individual forward observers can make the highest and most meaningful contribution to an artillery battalion's performance. They are the eyes and the ears of the battalion. The gunners, in particular, can be motivated to high performance when they are informed that the target is real, live, and significant. Gunners eventually establish their own opinion ratings of the lieutenants who are responsible for the observed fire missions. The forward observer's diligence, along with not unnecessarily exposing his driver and radio operator to enemy fire, are also topics for the battery grape vine. Forward observing is the primary testing ground for artillery lieutenants. The basis for grade is contribution to the battery's role. This does not necessarily mean self-endangerment. It does mean the use of logic, common sense, and an understanding of the enemy and his tactics. Your overall contribution as a forward observer is also enhanced by being in the right place at the right time, if you can so arrange it.

Lieutenant Denton had a cool head on him. I respected him a lot (Read 1993).

The unmapped wooden bridge

About 17 miles beyond the Seine River crossing there was an unfordable creek. Retreating German forces had blown the bridges. I was on foot with the front-line infantry and we crossed with a few small boats. I realized my jeep would be delayed pending construction of a new bridge. And, I knew our entire battalion was on the road, just to the rear of a tank battalion, a few miles behind us.

An elderly Frenchman flagged me down as I reached the far side. I will never know why he selected me, but I could easily understand that he was excited and wanted my full attention. He led the way along the creek bank until there appeared in view a fully intact rickety wooden bridge. I estimated it would easily carry jeeps and probably a truck with a towed howitzer. It was time for one more decision. The infantry column could go on without me, and I would eventually catch up to them in my jeep. Jacques Linn and my jeep were not far from the intact bridge. Linn raced back in our jeep to the roadside column, pulled our three howitzer batteries out of column, and guided them down to where I was stalling off a tank company. I had been in the Army for five years and had learned that putting on an air of authority amongst strangers would accomplish wonders. There was an Armored Force Major and several Captains, but I convinced them we had the priority. Linn was superb at following directions, never allowing anything to prevent him from complying. It was obvious to me the bridge would not support tanks. I assumed command of the situation and allowed only one truck and howitzer on the bridge at a time. Our three batteries made the crossing. Highhedge (the code name for our battalion) became the most advanced artillery battalion in that sector. Then, the tanks peristed in testing the bridge despite my objection. The first tank broke through and the bridge became useless. No other vehicles, nor artillery, would advance beyond this creek for 20 hours. But, our advancing infantry was now closely followed by one battalion of artillery.

Neuilly

I was with the advancing Infantry as we approached Neuilly. Suddenly we experienced German anti-tank fire from the vicinity of a prosperous French farm about 2,000 feet across a deep ravine. The column was halted. One of our tanks was moved up and commenced direct fire. The French farmhouse started burning and then we realized the enemy guns were in an emplacement which was 100 feet away from the house. We re-directed our deadly tank fire and a few more German soldiers had given their lives for "Lebensraum." A number of French civilians poured out from the basement of the house and tried to suppress the house fire. Our column was moving again with some remorse, but we could not take time out to help the French firefighters. Such is war (*C'est la guerre* as the French would say).

The German OKW — Wehrmacht Headquarters — 29 September

The German catastrophe at Falaise resulted in General Model succeeding von Kluge as commanding officer of Army Group B, and the western front.

Major Percy E. Schramm, was the officer in charge of the war diary of the Wehrmacht's (regular German army) Operations Staff. After the war he became a Professor at the University of Goettingen. The war diaries and files of the Wehrmacht were systematically destroyed at the end of the war. However, Major Schramm, in collaboration with others and from memory, re-wrote some of the Wehrmacht's diary (Schramm 1946).

The losses suffered by the Germans since the invasion were enormous. On 29 September 1944, the German High Command War Diary listed 516,900 killed and captured plus a further 95,000 who had been bypassed and were defending coastal fortresses on the Atlantic.

Earlier in September, the German High Command War Diary listed their western front strength as 48 infantry divisions, 15 Panzer divisions, and 4 Panzer brigades. Of these, 9 infantry and 2 Panzer divisions were reorganizing. Furthermore, 14 infantry and 7 Panzer divisons were battle-weary. At the same time they estimated the Allied strength on the western front as 54 divisions. Thus about 60 German divisions opposed 54 Allied divisions (Schramm 1946: p. 28-29).

German intelligence organizations, and this applied particularly to the *Fremde Heere West,* usually tended to exaggerate the strength of Anglo-American forces threatening to invade Fortress Europe (Seaton 1981: p. 69). In this case, whereas the Germans estimated there were 54 Allied Divisions, there actually were 49 divisions.

Belgium

The first two days in September it appeared German resistance had crumbled. Infantry troops were packed onto tank decks and the race for Belgium was on. The tanks were the 2nd Armored Division's. I finally got off my feet and rode along in my jeep. The advance used every possible road and so there were numerous advancing columns within the sector assigned to the 30th Infantry Division and the 2nd Armored Division.

There was a deep-seated reason for the rush to reach Tournai, Belgium. General Bradley, Commanding the 12th Army Group, had repeatedly objected to the use of airborne troops during a pursuit. Yet, Eisenhower had scheduled a large airborne drop near Tournai for September 3rd. Bradley wanted to be in Tournai in time to nullify the need for the airbrne drop. Incidentally, Tournai was inside the zone assigned to the British Army (Blumenson 1963).

Our Artillery Battalion raced 125 miles, from the last gun positions at Neuilly, in two days time. We were usually following one of the armored columns. Meals were cooked in a moving mess truck, and served along the road. Frequently, three or four men would fill up their mess kits and then the word would come to move out. There would be additional stops for those men who had not eaten that particular meal. Night driving was blackout with only "cat-eye" lights. Some stops were for an hour or so. Sleepy troops used this all to advantage and slept along the side of the road, leaving one man alert to waken them when the column moved again. No one was left behind. No one wanted to be left behind. This was history in the making.

I dropped back to rejoin our battalion late on the 2nd of September. As we approached Belgium, roads became more numerous and the advancing columns further divided like a river delta. The 283rd Field Artillery Battalion was soon on its own. Shortly after dark we reached the town of St. Maur, in Belgium, just east of Tournai. The town was uneasily quiet, all homes being heavily shuttered. Eventually, a shutter cracked open and some one shreiked "Amis" or Americans. That did it. The whole town came alive. Dark and misty as it was, deliriously happy throngs of Belgium people broke loose with what must be their record-setting celebration. Those church bells were ringing again. The 2nd Armored Division, and elements of the 30th Infantry Division, liberated Tournai well before midnight thus cancelling the planned airdrop. And once again, Montgomery protested the American incursion into his terrain (Blumenson, 1963:p. 388-9).

My forward observer team took over a tavern where the owners broke out Benedictine, a sweet liquor. They had saved it for this occasion. There was a light rain so the team and I simply slept on the barroom floor. We realized something was wrong when one of our night guards excitedly reported he had attempted to awaken his relief and the man had responded in German language.

> After we set up our howitzers, a few of us from "A" Battery
> bedded down in a nearby barn. Early the next morning we found
> that German troops were also bedded down in the same barn.
> We got the jump on them (Kolakowski 1994).

Church bells may well have rung all night. They were ringing away the next morning which happened to be Sunday. The Mayor planned special Church services. Some of us were on our way to Church when spotty, poorly aimed sniper fire commenced. It was difficult, as usual, to detect the source. We had an attached 40mmm anti-aircraft team with us and they opened up on the third floor attic window of one house. It turned out the only occupant of the room was a baby who, most fortunately, was not hurt at all. In the meanwhile, Lieutenant Bill Grohne of Battery "A" had climbed a farm silo only to determine the German fire was from an adjacent woods. Battery "A" moved howitzers into position and opened direct fire at a range of only 500 yards (Reber 1993). Soon a white flag appeared and 120 German troops surrendered. It turned out their captain would not allow surrender, but he was killed by one of our early artillery shells.

Later that Sunday the Mayor gave a speech in which he stated the 283rd Field Artillery Battalion's liberation of St. Maur would be memorialized in the town square.

The advance party heads for Liege

The battalion moved into bivouac at Fontency, Belgium on September 6. The next day the battalion continued the drive toward Germany and travelled 60 miles to a bivouac area at Gaillemorde, Belgium. Cities passed through included Tournai, Bruyelle, Camiley, and Nivelles. I was assigned to an advance party of the 30th Division which was going to race ahead towards Namur and Liege. German opposition had crumbled. We moved out — I in my jeep with trailer well stocked with rations.

Red Ball Express

Unbeknownst to me, the battalion was reassigned to a new mission a few hours after I left. The battalion was to go into encampment at Ohain, France, unload their 2 1/2 ton trucks, and join in what became known as the Red Ball Express. The American supply lines had become so long, from the Atlantic Ocean ports to the front line, that severe measures had to be taken. For the next few weeks our truck drivers, assistants, radio operators, and command vehicles shuttled supplies forward, driving both night and day. Certain roads were reserved exclusively for the Red Ball Express. One of our soldiers, encountering another convoy of troops, enquired as to what they were hauling. "Gallon cans of sliced pineapple" came the reply. "Throw me one?" It had been so long since he had tasted any that, despite his protesting stomach, he ate it all (Pensock 1993). In deference to German aircraft, night driving was with no lights except the cat-eye lights used in blackout driving.

Individual replacements

Individual replacements commenced arriving at about this time. Few were from Pennsylvania. They had been through the gigantic training centers in the states, shipped overseas with few, if any, acquaintances, and processed through replacement depots in France. They were naturally sceptical about their new assignments. They had no sense of inferiority but they had anxiety to perform well and to be accepted by the troops. Any fears they had were quickly dispelled. Yes, I suppose there were a few exaggerated war stories, but everyone who joined "C" Battery, and probably the Battalion, were swiftly embodied into our team.

There were adverse reports concerning the efficient operation of the replacement depots. At one point, the Inspector General assigned a young officer to go undercover, through a replacement depot, as an enlisted man. The resulting report led to beneficial modifications in the system. I vaguely recollect reading of this undercover work in the "Stars and Stripes", our Armed Forces daily.

Newly arrived artillery battalions

New battalions of U.S. heavy artillery were also arriving in France. Some, activated in early 1944 in the States, were in France late in 1944.

> After the 1944 debacle at Cassino, Italy, which confirmed aerial bombardment could not substitute for heavy artillery, the War Department surpassed the highest proposals made by the Army Ground Forces at earlier dates. Increases were made chiefly in 8-inch and 240mm howitzer battalions as a result of the April 1944 Lucas Review Board (Stanton 1984:p.29).

Namur, Belgium

In the meanwhile, the advance party I was with arrived in Namur where, it became apparent that there was no front line. We had by-passed numerous German troops along the way. The colonel in charge of the advance party told me the 283rd Field Artillery Battalion was no longer attached to the 30th Division, and I was released from his force. He had no other information.

Our team moved to high ground, above a wooded area, and attempted radio contact with our Battalion. In the meanwhile, the remainder of the advance party moved out of Namur toward Liege. Linn established good relations with a group of teenagers from Namur. We proceeded to swap our canned rations for fresh eggs and bread and lived fairly well for a few days. We relied on the townspeople to alert us if Germans were spotted. None were. Eventually we did make radio contact and headed towards Ohain. We used caution and queried people before moving into any town. Fortunately, no Germans were encountered and we arrived in Ohain. Martin Williamson, "C" Battery Commander, was on the Red Ball express, but someone had looked out for me. They had reserved a comfortable bedroom for me in a French home.

Ohain, France

There were developments in Ohain worth noting. Our kitchens had assembled in the school house yard, and were serving fairly good meals to the few troops who had not gone on the Red Ball Express. Our cooks had learned to mix powdered eggs into a batter and fry them like hotcakes. There were other innovations. The problem was that our troops were reluctant to eat in view of so many very skinny and hungry French children. A deal was struck with the town mayor. All left over food, scraps, and even used coffee grounds would be given to his representatives for distribution. He, in turn, would keep the children away from the troops at meal time. My host family had heard all about me before I arrived. They had planned a big meal for me, based on a roast of horse meat. Beef was not to be had.

The Northern France Campaign had ended. Eventually, we were all awarded a star on our European Campaign Ribbon indicating participation in one campaign. Three more stars would be added. And, in 1950-1951 I would garner an amazing six campaign stars in the Korean War. Survivors of frontline combat in ten military campaigns are scarce.

Awards for heroism were stringently awarded in World War II. In this one campaign, fifteen bronze stars were awarded to battalion members. There were no higher awards. Numerous wounded and killed were awarded the Purple Heart.

The Battalion went back into action in early October. The Campaign for the Rhineland would initially employ the 283rd Field Artillery as part of a small force screening the American left flank.

Chapter 3
THE CAMPAIGN FOR THE
RHINELAND AND THE BIG PAYOFF

Limburg Province of Holland — 1944

One-third of Limburg province in Holland was liberated by Allied Forces in Mid-September (van den Berk 1993). Limburg is the narrow sector of southeast Holland, between Germany and Belgium. The U.S. Second Armored Division, under XIX Corps, of 1st Army drove eastward across the Meuse. On the 18th of September they had driven across Holland into Germany at the town of Wehr, which is four kilometers east of Sittard. Major General Ernest Harmon commanded the 2nd Armored Division at the time. In his biography *Combat Commander* he writes:

> My first sight of Germany was this tiny town (Wehr). Not a soul was on the streets. All doors and ground-level windows were locked and barred, and from every second-floor window hung a white sheet that meant surrender. The German army had fallen back to make its defense at the Siegfried Line, which hugged the east bank of the Wurm River in that part of the country (Harmon 1970).

The German Town of Millen (Map # 2) and neighboring Millen Castle, in Holland, were liberated on 19th of September 1944 (Heemkunde Vereniging Nieuwstadt 1993). Jacques Briels, a 19-year old Dutch youth, remembers their tank liberators as being from the 2nd Armored Division, and one of their third battalions. Heavy German artillery fire immediately followed the liberation (Briels, 1993). Nieuwstadt (one km. northwest of Millen) and Holtum (three km. northwest of Nieuwstadt) were liberated by the 29th Infantry Division and the 113th Cavalry Group (van den Berk, 1993).

The U.S. Ninth Army

In early October there was a new grouping of American Armies. Ninth Army, which had been in Brittany, was slated to be inserted between First and Third Armies. The 10th of October plans abruptly changed. Ninth Army was assigned the zone north of First Army, instead of south of First Army. Thus, Ninth Army would have the British on its left and First Army on its right. This dramatic shifting of Armies caused a delay from the 5th to the 12th of October for our Battalion and probably many other units. Both Ninth and First Armies were in the 12th Army Group, under General Bradley.

Reasons for switching 9th Army to the north

General Eisenhower and the Supreme Headquarters Allied Expeditionary Force (SHAEF) became established in France on 1 September 1944. Thereafter, General Montgomery commanded only the 21st Army Group which was comprised primarily by 15 British and Canadian divisions. They had landed in Normandy at the most easterly of the invasion beaches. Thereafter, in the movement north into Belgium the 21st Army Group was closest to the Atlantic Ocean and thus continually on the left of the Allied Forces (Map # 4). General Montgomery repeatedly requested General Eisenhower to assign some of the American Divisions to him so that he would have a larger command to engage larger objectives. General Bradley, commanding the 12th Army Group, was not supportive of Montgomery's requests. Yet, Bradley realized that Eisenhower would eventually give in to Montgomery and possibly assign an entire American Army to Montgomery. This realization came to the fore at this time, and it seemed prudent to Bradley to lose the incoming 9th American Army to Montgomery rather than the battle-experienced American 1st Army. It also seemed that General Simpson, commanding 9th Army, was one of the few American army commanders who could get along with Montgomery. Hence the switch so that 9th Army would be closest to the British. As matters turned out, it was to be about five months before 9th Army was assigned to the British 21st Army Group for an offensive operation. Montgomery was also assigned 9th Army during the Ardennes German Offensive (Whitaker, Whitaker 1989: pp 14-15).

In deference to the shortage of transportation, instead of physically shifting troop units, there was a wholesale trading of units. The XIX Corps, including the 283rd Field Artillery Battalion, became a part of Ninth Army. Lieutenant General Simpson officially opened Ninth Army Command Post at Maastricht on October 22nd. He was a tall, lean, West Point 1909 graduate who was popular with the troops.

The left flank of American forces

Our 283rd Field Artillery Battalion departed Ohain, France, on 5 October and travelled 123 miles to Teuvan, Belgium where we went into bivouac awaiting new orders. This delay resulted from the aforementioned "switch".

On 13 October, 1944 we drove 29 miles, passing through Valkenburg and Sittard, to our new gun positions at Guttecoven, Holland. Our battalion was assigned as direct support of the 125th Cavalry Squadron (Mechanized) on a screening mission covering the left flank of American forces. The assigned frontage of 10,000 yards was exceptionally large for a squadron. Thus, the squadron occupied a few strong points and periodically conducted patrols between them. The British Second Army, commanded by Lieutenant General Dempsey, was further left.

The 125th Cavalry Squadron was under the 113th Cavalry Group. The Group was in XIX Corps, commanded by Major General Raymond S. McLain. Actually the Cavalry Group was operating north of Ninth Army's northern boundary, but the British did not have any available units to replace it. British forces assumed responsibility for Holtum and other parts of Limburg Province in early December.

Our 283rd Field Artillery Battalion headquarters command post was at Einighausen. Battery "B" was also at Einighausen. Battery "C " was at Guttecowen. These towns are in the Southern Limburg Plateau, west of Sittard (Map No. 2).

The artillery ammunition shortage

American Armies were still short on ammunition because the operating port facilities were so far from the front line. For a while we were limited to two rounds per howitzer per day, unless we were attacked. A later restriction limited our firing to only if we were attacked.

> Supplies from the rear couldn't be sent to our battalion fast enough. One night our ammunition train was about three miles from the outskirts of Brussels when we ran out of gasoline. We pulled off on the side of the road to allow a British convoy to pass. The officer in the lead jeep stopped to ask "Is this the road to Brussels?" When we stated "yes", he responded "Has it been taken yet?" (Parker 1994).

The American 1944 Presidential election

The American Presidential election was only weeks away. I suppose it was felt that most soldiers would vote for President Franklin D. Roosevelt. In any event there was a concerted drive on, for those who were 21 years of age or older, to apply for absentee ballots. I qualified and so I did vote this one time for President Roosevelt. His opponent was Thomas E. Dewey. In 1948, Dewey was again the opponent, this time to President Truman.

At the enemy's headquarters, and unbeknownst to us at the time

When Antwerp fell it appeared the German 15th Army might be trapped, but by the 23rd of September 82,000 German troops and 580 artillery guns had been evacuated to northern Holland and eventually to Germany via a northerly route (Schramm 1946:p.35). It was a difficult series of night-time moves because of the need to conceal it from Allied intelligence. General Gustav von Zangen was one of the old "Prussian Style" officers who was devoted to Adolph Hitler, although like most German generals he was very critical of Hitler when the war ended. Von Zangen's new mission was to relieve the 5th Panzer Army which had been defending a northern part of the German Westwall opposite the Limburg Province of Holland. Their sector included Millen, on the Dutch German border.

> Von Zangen's 15th German Army Chief of Staff was Obst iG von Kahlden; his IA was ObstiG Metzke; his OQu was Genmaj Knauff; and his senior artillery commander was Genlt Burdach (Schramm 1946: p.258).

In early October the staff of the German Fifth Panzer Army was temporarily assigned responsibility for commanding the defenses of Aachen. It was an important German city, having been at one time the nucleus of what was now the German Reich. Desperate measures were needed to prevent American forces from capturing their first large German city — and Aachen at that. General von Manteuffel and his staff of the 5th Panzer Army were withdrawn from Army Group G on 15 October and moved into the Aachen area (German Military Studies; p.40). The components of the Army remained in place. Advance elements of the staff of the 15th German Army were already in the zone. They were in the process of relieving the Fifth Panzer Army. Temporarily, they assumed staff responsibilities for both armies.

Our observation posts at Roosteren and Nieuwstadt

One of our Battalion's observation posts was in the church tower at the east edge of Roosteren. This is at the very top center of Map # 2, between the Maas River and the Juliana Canal. The French and Belgium Meuse River is known as the Maas River in Holland. They are one and the same. Roosteren was nine kilometers north of our gun positions at Guttekoven and Einighausen. Fire missions directed from this observation post were straining the range of our guns and always required charge 7. The Roosteren observation post was shared by both "A' and "B" Batteries. Lieutenant Fred Whitt wrote about it in his unofficial diary. Lieutenant Bill Grohne recollects it well because the drive north to the post was always a scary affair. There were no friendly troops along most of the distance. The 125th Cavalry did have an outpost in Roosteren. German patrols operated in between our outposts. The front line in the Roosteren area was the canal. Allied forces were exceptionally vulnerable to any German attack in this area. Nevertheless, our artillery fire directed from this observation post dampened some German thoughts of attacks (Grohne 1993).

Another of our Battalion's observation posts in October was in the church tower at Nieuwstadt. Lieutenant Fred Whitt, of "B" Battery, and possibly others, adjusted artillery fire from this observation post (Whitt, 1945).

Knights of Millen Castle

Battery "C" was in direct support of Troop A which was located in Holland on the west edge of the German town, Millen. The German town of Millen had been liberated as aforementioned. One of the Cavalry outposts was at the Manor House, directly on the international boundary. I located my observation post in a tower which was a remnant of the mediaeval Millen Castle of the 12th century.

In the year 1118 the powerful von Millen family included the Bishop of Cologne. The castle was constructed to protect the family and their assets from various forces including the powers of the Rhineland. The von Millen family and the Knights of Millen were able to withstand these repeated onslaughts. Archeological explorations in 1964 revealed the castle was about 55 meters in diameter (Eussen 1967). Millen Castle, excepting the "Dongeon Tower", was torn down in 1650 when the Manor House was constructed.

> Millen Castle, which is on Dutch territory, was separated from the German village of Millen when frontier corrections took place in 1815, pursuant to the Congress of Vienna. The Knights of Millen lived in the castle in the 12th and 13th centuries. At the end of the 14th century various parties fought for the possession of Millen. It was the Duke of Brabant who finally won the battle. In 1467 the castle came into the possession of Maria, the sister of the prince bishop of Liege, who was also the Lord of Millen. (Strik-Zijlstra, 1993).

The duchy of Brabant existed from 1190 to 1430 when it was united with the duchy of Burgundy. The reader will find more detail about the Duke of Brabant in most encylcopedias.

The Dongeon Tower, a remnant of the Castle

Its (Millen Castle tower) height is about 19 meters, its inner-circle is about 4.6 meters and the thickness of the walls is 2.5 meters. The tower has three floors. The entrance downstairs has been made afterwards. Originally this tower floor was meant to be a gaol (jail) and only accessible from above. The arch of this lower floor has been mainly destroyed. The original entrance on the second floor was attainable by a wooden bridge from the castle walls. From this second floor a staircase, built inside the walls, led to the summit at about 19 meters (Heemkunde Vereniging Nieuwstadt, 1993).

In 1944 the walls were covered with vines. There was brush vegetation on the top. From a distance, the tower blended in with the surrounding trees. Inside the tower there was a sixteen-foot retractable ladder which led to the winding path-like stairwell aforementioned. There were two places where there was no inside wall, allowing the climber to view the possible drop should he slip. The ascent to the tower roof was highly challenging, but then we had no official visitors.

The top floor has three apertures. The northeast view, from either the top floor or the roof, covered some of Millen, a large German farm field of about 40 hectares, a small isolated island of trees with a house in the northern part of the field, and a wooded area to the north. The terrain was flat except that Millen Castle is slightly higher ground. Range to the island of trees was about 1,000 meters from Millen Castle.

In the 17th Century a grey-stone castle-like house was built 80 meters east of Millen Castle, or more precisely, the tower remnant of Millen Castle. In order to differentiate the house from the tower remnant of Millen Castle, I am referring to the house as the "manor house". It is in Holland, directly inside the Dutch border. The actual border in this region is Red Brook. In order to reach the manor house, or the castle, by vehicle one must first cross over into Germany for a few meters.

A 1941 German map of the terrain (Map 2) shows the castle, manor house, and Red (Rode) Brook. The woodland clearing where we had our encounter with the German officers is marked with an "X". The cross-section grids are at one kilometer spacing.

The terrain still being contested

After the Second World War the Dutch government demanded compensation from Germany for destruction of Dutch Atlantic-Ocean dikes. The Dutch threatened to annex German territory if Germany failed to comply. In 1949, the boundary was shifted five kilometers easterly. Holland was given 70 square kilometers containing 10,000 inhabitants. The town of Millen was included in the ceded terrain. "The frontier corrections did not benefit Dutch-German relations, and from the mid-fifties onwards negotiations took place with regard to handing back the territory" (Strik-Zijlstra, 1993). The areas were eventually returned to Germany in exchange for monetary compensation.

Germans, living in Millen, were evacuated by train in early September of 1944 and sent deeper into Germany. When we arrived, Millen was deserted and considered "no-man's land" except for the area around the 125th Cavalry's outpost. On the Dutch side of the border people continued with their apple harvest and other chores during October. They produced the best eating apples I found in Europe. Two months later, after our departure, the British forces evacuated the Dutch residents from towns along the front line such as Holtum and Nieuwstadt (van den Berk 1993).

I ventured in to Millen on occasion to use the church tower as an alternate observation post. This was hazardous, even in daytime. German patrols also ventured into Millen. German forces never detected our use of Millen Castle as an observation post, because from a distance it blended into the surrounding vegetation. The castle was indicated on German maps.

Observation from the Dongeon Tower

I used the small house, at the north end of the field, as a check point and registered a few rounds when I first occupied the castle. Our observation was enhanced with binoculars and a small twenty-power telescope. It was an easy registration because we had 1:25,000 black and white maps which were so good you could even select a certain tree in an apple orchard. Our Fire Direction Center used the same set of maps to calculate the data to send to gun positions. Observers only had to report the coordinates, and then adjust as necessary. Wind, temperature, and weather conditions usually created a need for adjustment.

The small house, in the island of trees, was used by German troops as an outpost. Frequently we would see German soldiers basking in the autumn sunshine on the southwest side of the house. They seemed to know about our firing restriction. The Germans were living in the cellar of the partially destroyed house. Doubtlessly, it had a stone-arched ceiling and was relatively safe from artillery fire.

The Dutch Resistance lad

A 19-year old Dutch lad (Jacques Briels, now living in Utah) was most helpful to us. Jacques was well educated, and fluent in English. He provided our patrols with information on the terrain, and provided start-off guidance to a few patrols. It was unusual to find any healthy 19-year old European male not in uniform, especially when he was fluent in English. Supposedly all eighteen-year old Dutch boys were required to serve one year in the German labor service organization. Jacques, and a close friend, received their notification to report for labor service (Jacques still has his notification card and recently gave me a copy). His friend felt obliged to report or otherwise his family would have suffered severe punishment. Jacques, however, moved off to a farm seventy kilometers north and hid out until the region was liberated. His friend eventually returned with some hair-raising experiences. Jacques' older brother was in German labor service in the Stuttgart area. After undergoing one severe bombing raid, his brother took off for Holland and eventually hid out with Jacques. The former Hitler youth from Millen were being slaughtered. The 12,000 Dutch lads who had sworn allegiance to Adolph Schiklegruber, and joined the German army, also were being slaughtered — probably on the eastern front. There were more than 300,000 Dutch people enslaved in Germany. Most Dutch Jews were killed or in German detention centers. Thousands of Dutch people had been ruthlessly killed, in hundreds of atrocities within the Netherlands, by the German occupation forces. These murders were German retaliation for Dutch resistance and especially Dutch opposition to German treatment of Jews (Maass 1970). This one Dutch lad was a survivor.

A forward observer, secured by the Dongeon Tower

We were at this location from 13 October until 6 November 1944. My team and I occupied the castle on a cycle of 48 hours and then were relieved by our battery's other observer team. My radio operator and I would live in the castle, pulling up the ladder at night. The castle was isolated and vulnerable to German patrols. We did keep a stock of grenades on hand to welcome any Germans venturing into the castle. The entranceway had no door at all. Our jeep and driver were in the manor-house barns along with the 125th Cavalry's drivers and vehicles. I appeared younger than my actual years and as aforementioned did not wear my officer insignia. Linn had difficulty convincing the cavalry troopers that I was an officer. He did succeed in filling them full of beans about our exploits.

Our Fire Direction Center had probably tagged me as a judicious fire-mission requester. The Battalion fired 45,361 artillery shells in Europe. I have no record of the artillery rounds that I shot at Germans troops. It was about 800 rounds. Generally I fired only when I had a definite target. I was not one of those observers who fired at every suspected target, even though they were only doing what the 'book' says to do in such situations. I have always been frugal in nature as a result of growing up in the great depression. Fire direction center soldiers eventually had their own assessment of each forward observer. My past discretion was to pay dividends at Millen.

G.I.'s of Millen Castle butcher the Teutonic butchers

George W. Hoy, now living in Lock Haven, Pennsylvania, and I were on the roof of the castle on Sunday morning, 21 October 1944. It happened to be George's birthday which facilitated later recall of the event. George called my attention to some movement north of Millen. I observed a group of German soldiers in a small clearing in the woods on my left (Viehweide — north of Millen and just east of Nieuwstadt). Their numbers grew until there were 35 and amazingly all were Wehrmacht officers. With the exception of two, they were dressed in their jack-boots, soft hats, and be-medalled tunics. The other two were helmeted and in combat uniform. Several had maps and apparently were discussing a crossing of Roode Brook. These German officers were planning some operation. It seemed as though they were becoming familiar with new terrain to which they had been assigned. They were not attacking us. They had no idea that they could be observed because the top of the castle appeared to be just some more tree tops. The view from Millen Church tower did not include their location. They surely knew this portion of the front line was thinly held with only a few strong points, and that intervals between strong points were periodically patrolled. Captain Jaydee I. Dodson, the fire-direction center duty officer, reminded me of the policy against firing unless we were attacked. He also knew that I only called for fire when I had a real-live target. And, this Sunday morning I was convincing. This was the opportunity I had dreamed of since 1933. There was no way I was going to let them go. I had to be convincing. My imagination assured me these were the devil's own. Of course, I really did not know them nor could I recognize any SS insignia. I had been an avid reader for at least a dozen years, and I felt the German Officer Corps was dominantly responsible for this war which, even at that time, had caused world-wide deaths and maiming of many millions of people. Within seven more months they would have caused the obliteration of one entire age-class of German men, by allowing the war to futilely drag on. The 20 July 1944 attempt on Hitler's life was too little and far too late. Youthful German troops were also the devil's own, but it was the German Officer Corps who had lured and led the young men into Hell. I was, and am, also convinced that most of the thousands of incidents where innocent men, women, and children had been slaughtered to set an example for potential resistors to German domination, were ordered by the German officer corps. Numerous German officers (of field grade and higher) survived the war, went unpunished, and are drawing liberal pensions (Diehl 1993). Now, we had 35 of them in my grasp. These devils were not going to survive the day.

Jaydee typified the American reserve officer, and used his own judgement despite orders. Let me digress to mention that I had never met a regular army officer, except Brigadier General John Lewis, up until the end of the war. Our fighting ground forces in World War II were led, in the lower echelons, by our civilian army, reserve officers, dedicated to winning the the war and rejoining their families. Few, if any, had strong career or promotion aspirations. Time and time again, they acted contrary to orders when they felt their course of action would further our victory. The Battalion staff of the 283rd Field Artillery Battalion typified the American:

> They were practical inventive, people who made wide use of expedients. Initiative, individualism, fortitude, and exuberance dominated the scene (Turner, 1920).

German troops were noted for their obedience. There is little doubt, that in a similar case, no German officer at battalion level would have dared to approve the fire mission. Possibly these 35 German officers were relying on our conformance to their obedience standards.

A normal artillery adjustment on the target was not feasible. It would have frightened them away. We had to chance it with a battalion salvo (meaning one round fired simultaneously from each of 12 howitzers). Our three firing batteries swung into action. The troops were informed of the unusual nature of the target. The battalion time-on-target twelve-round volley was short. The

Germans were turning to leave and bantering with each other to show their bravery. Possibly they were still under the belief that their position could not be observed. I hastily made a range adjustment. Our eight-man fire direction center, half from the Kentucky cadre and the remainder from Pennsylvania, swiftly converted my range adjustment to fire commands. The fire commands included a slight change in direction for some batteries along with raising the angle of the guns slightly. The commands were transmitted to each of the batteries by the fire direction center man who was in continual contact with each of the batteries during a fire mission. Our Pennsylvania gunners reacted with speed and accuracy which would have rivaled our computers of the 1990's. Less than one minute from the first salvo the second salvo caught them. Time was of the essence. Thirty seconds more would have meant some might have escaped. Our fire direction center and our gunners evidenced the highest possible degree of proficiency. Three artillery shells hit trees, creating low airbursts which were deadly. One round struck in the center of the group. Five survivors headed east, only to run into our next near-instantaneous salvo, for we had anticipated their reaction. Jaydee interrupted to say Corps had heard us firing and ordered us to stop. The order was of no consequence. In less than five minutes time I had more than paid off the costs of my five years in training. Millen Castle became the capstone of my career. Our gunners and fire direction crew were ecstatic.

The incredibly short time span between salvos must have dumbfounded the Germans. No other army had developed procedures which facilitated this near-instantaneous reaction to fire commands. The men in our battalion could not be matched in these aspects. This incident reflected most favorably on the upbringing of these Pennsylvanians. This case history is evidence that field artillery gunners are far more than laborers.

Three hours later, a German medic crept into the scene. I cannot explain the lengthy delay, but I know it was afternoon. We did not fire. Then followed an ambulance, and then more. We witnessed the removal of bodies all afternoon, at the same time conveying our observations to fire direction center. They forwarded the information to all three battery gun positions. The delayed evacuation probably meant that each removal was a dead body. Thus we had a good estimate of the killed. Somehow, I wished those children, whom these scoundrels had marched off to death camps, could have been with me. Perhaps the Knights of Millen Castle were there in spirit.

Most Dutch residents of the area sought refuge in their cellars when the firing started. They well remember the incident because artillery shelling had been a rare event, in recent weeks, due to our ammunition shortage.

Only two people saw the dead or wounded officers being transported and wondered why no weapons were fired by infantry, artillery, or tanks while an observer plane was in the air (van den Berk, 1993 interviews of residents).

There was an American light observation plane in the area at the time. The Dutch believed this was the mode of observation for the fire mission. Then, they could not understand why the plane would permit unhampered transport of the dead and wounded. The Dutch never realized the artillery adjustment emanated from Millen Castle, until our 1993 correspondence. They did not favor allowing the evacuation of the dead to proceed (van den Berk 1993).

At the time, I wondered what the German officers were up to, and who they were. It would have been possible to retrieve a few bodies, immediately following our shoot, but Captain Dale D. Stithem of "A" troop, 125th Cavalry had no interest in a day-time patrol. He did not keep a diary and has nothing to add to this report (Stithem 1993). In a recent fire fight the 125th Cavalry had lost three officers and 12 men (Eagen 1975). Their reluctance to engage in a daytime patrol is understandable.

Wehrmacht records were systematically destroyed at the end of the war. As aforementioned, the offical diary keeper, Major Schramm, rewrote some of the information from memory and other sources after the war (Schramm 1946). Schramm did include overall casualties in his 1946 writing. But casualties from any single event were not recallable. General von Zangen wrote his after-action report in 1946, but his report commences about a week after our encouter with the 35 German officers.

We did know the German 176th Infantry Division, 183rd Infantry Division, and 9th Armored Division were opposing us in this sector. The German 15th Armored Infantry Division was in the area, but the location was unknown. And, we thought these units were part of the German 5th Panzer Army.

The last resistance at Aachen ended at the very same time that 35 German officers were being slaughtered near Millen. This loss of the first large German city to American forces was a critical blow to all Germans, and particularly their armed forces.

Unbeknownst to us at the time

We were not aware of the new role of the German 15th Army. General Model had withdrawn the 15th Army along the coast past Boulogne, Dunkirk, and Calais to the banks of the Schelde Estuary in Holland. Then he had ordered General von Zangen, who commanded the 15th Army, to break through to the east. Their advance party arrived in our sector in October. By the 15th of November the 15th Army had replaced the German Fifth Panzer Army which had been opposing us in our sector. The Fifth Panzer Army was moved south to assume the main attack in the December Ardennes Offensive. German plans were for their 15th Army, on the north flank of the penetration, to conduct a subsidiary break through from Millen to Maastricht where the Americans had immense supply bases.

This Millen Castle incident could well have been the intelligence tip off for the forthcoming German Ardennes offensive. It wasn't.

The 35 German officers apparently were part of the advance staff of the 15th Army. They were reconnoitering the attack route shown on the map opposite page 545 in Wilmot's *The Struggle for Europe*. The sector, immediately north of Millen, had been a dominant salient in the May 1940 assault on Netherlands (see Chapter 1).

> Another obstacle (to the American Rhine River Crossing) was
> the powerful German position from Roermond to Duren on the
> German side of the (Dutch) border. This strong bridgehead was
> a constant threat in the direction of Maastricht and the impor-
> tant Allied base at Liege (Maass, 1970. p. 207).

And, another quote which could shed some light on the purpose of the 35 German officers at Millen (Millen is midway between Geilenkirchen and Maeseyck).

> In view of the German salient on the thinly held Army north
> flank from Geilenkirchen area to Maeseyck and the road net
> favorable to a possible German attack in the general directions
> Heinsberg-Sittard-Gangelt, the Army Commander located a
> small reserve in the area south of Sittard (Infantry Journal. 1947.
> p. 73).

German General Model, and other top ranking German generals had been urging "The Small Solution" as an alternative to the the forthcoming December Ardennes Offensive. The "small solution" was the plan for a German attack through Limburg Province to capture the immense American supplies at Maastricht. In a meeting with Hitler, on November 3rd, Rundstedt and Model again proposed to launch one drive from the north (the small solution) (Merriam, p. 28). As late as December 2nd, Model made a last moving plea for the "smaller solution". Hitler could not be moved from his dream. The Ardennes

Offensive would proceed as planned. If it were successful, the 'small solution' would proceed as a secondary effort.

The Germans seek revenge

My replacement forward observer jestingly said he was unhappy with me because he felt the Germans would try to locate our observation post and take revenge on him. He was an excellent forward observer and had undergone more than his share of German shelling. His jests, however, became reality. For the next few days the Germans repeatedly shelled the church tower in Millen. The Church was empty. German night patrol activity increased significantly. A few days later Linn and Hoy were returning to the battery along a dirt road near Millen Castle. They had replaced the sandbags on the floorboards of the jeep with armor plate. Sandbags were some protection from vehicular mines, but armor plate was better. Their right front wheel struck a mine which probably had been recently laid by a German patrol. Fortunately, thanks to the armor plate they were only shaken and not injured. The jeep was a total loss. Before this incident, Jack Briels, the Dutch lad, had alerted the 125th Cavalry to a stockpile of mines near the site but we were never notified.

At about the same time, back in "C" Battery's gun position on the edge of Guttecoven, Victor Krivitsky was attacked by a German patrol while he was on duty in a gun position at about midnight. The rest of his crew were sleeping nearby. There was a struggle and and a brief knife fight which woke the rest of the crew. The attackers got away, but Krivitsky had a very scary experience. Victor recovered in a matter of minutes and required no medical trauma treatment. I wonder if I would have done as well.

One of our men on guard duty proved his marksmanship:

> It was a cool, dark night and dry as we had not yet had a snow. Hearing a noise I shouted "Halt". The noise continued. "Halt" I shouted again. Still the noise continued. "Halt" and I fired into the darkness not once, not twice, but three rounds into the sound. Everything broke loose. Men came running. Then I realized that three shots were the standard call for help. They looked for my prey. In a few minutes an officer returned. "Good shot corporal, you shot a cat in the head" (Pensock 1993).

German forces did try one large scale attack on Holtum (Map 2) on the 28th of October. The 9th Panzer and 15th Panzer Grenadier Divisions were employed in a local attack operation conducted to relieve the hard pressed 15th Army (German Military Studies p. 43).

> In the night of 28/29 October 1944 in Holtum took a counter-attack place with disastrous consequences for the Germans, 250 casualties (van den Berk 1993).

Years later I learned that 12th Army Group's instruction to Ninth Army for the operations between October 4th and October 22nd included:

> "keep the situation fluid on their fronts at all times and to use every opportunity to kill Germans" (Infantry Journal. 1947. p. 62)

KIA, POW's and homefront shock

Lieutenant Jack Hare, of "B" Battery, three of his men, and two Dutch soldiers failed to return from a mission on the 27th of October. Accordingly, they were reported as missing in action. Months later, one of the men, Carl G. Devitz, was liberated from a German prison camp by the British Army and wrote his battery commander.

> We were on the way back to Base Camp when we struck a land mine and came under German small arms fire. I believe Lieutenant Hare suffered severe internal wounds, William Wagenman took two shots through the chest, and my right leg was shattered. The Germans hauled us in a wheel barrow and then on a one-horse two-wheel hay wagon to a German military hospital miles away. It was a difficult trip. Lieutenant Hare and I were lying side by side on this hay-wagon ride. He was unconscious most of the time and I couldn't ask him about his wounds. The floor of the hospital was filled with German wounded on stretchers. Soon, a German doctor told me "Your officer just died." (Devitz 1993).

William Wagenmann, another of the men who were with Lieutenant Hare, was liberated on the 13th of April 1945. He added to the above

> One of the Dutch boys was killed with injuries to the back of the head. I was shot after coming down out of the air. My wife was notified that I was missing in action on November 11, 1944.

She did not know anything more until she received a letter from me in early 1945. She then notified the Adjutant General and the Red Cross (Wagenmann 1993).

Carl Devitz's wife was notified he was missing in action. His distraught and worried wife did not receive the War Department's telegram, and the glorious news that Carl was alive, although a prisoner, until February 23, 1945. The four-month delay, in knowing if Carl was alive, was an extremely difficult period for her and other members of the family.

Home-front suffering was also experienced in Carysbook, Virginia, where Mrs. Perkins learned that her son Jack Hare was missing in action. Later she learned of Jack's death. She immediately commenced correspondence, and difficult war-time travel, to learn more of Jack's final hours. Her travel included a journey to my parents' home. Her later travel included a visit to query Carl Devitz while he was convalescing. After the war, in accord with her request, many of us visited with her. Included were William Wagenmann and his wife. Jack was an excellent young officer and a highly promising young American leader. The trauma shattered the life of his Mother. Some of us maintained contact with her at Christmas and Mother's Day until we were fully involved in the Korean War.

Life in Holland

Our mission in Holland was plagued by rain and mud. There were few paved roads, and many of the dirt roads were deeply rutted and muddy. The gun positions were a miserable place to be, with everything covered by oozy mud. I have not dwelt on this aspect. Suffice it to say this autumn and the ensuing winter were the worst weather conditions the Dutch had experienced in 25 years. November rainfall was more than double the normal.

We did enjoy weekly hot showers at the coal mine in Heerlen. The miners had a large, tile-floored, shower room. They did not have lockers, but were assigned a long rope on a pulley so they could tie their clothes on and haul them high into the air. Our troops enjoyed the chance to take off their damp boots and walk barefooted on warm tile floors. A few troops were granted 48-hour leaves to visit Heerlen, staying with Dutch families. Although there were no bathtubs, nor showers, in the neatly-kept homes each block of houses included a local bath house.

All personnel within our battalion shaved daily whenever possible. Our steel helmets, with their removable liner, were excellent basins for shaving and spot washing. The muddy terrain, along with the damp cold weather, complicated measures to maintain cleanliness. However, we really did very well.

86

Some of us in Service Battery were living in a huge kiln that had been used to fire brick. One day, Lieutenant Barrington came by and told me and three other men to "get lost for the day—and enjoy it". We went into the town of Valkenburg (Coincidentally our battery commander was Captain van Valkenburg). While walking down a street, a Dutch civilian stopped us and motioned for us to follow him. He took us to a cave outside of town where armed Dutch men were on guard. Inside the cave were more armed Dutch men. They showed us framed picture after framed picture. Tapestries were hanging from the walls. They were so proud of their concealing the National Art of Holland from the plundering Germans they wanted Americans to know about it (Parker 1994).

Direct support of the 29th Division

As soon as the Second British Army had sufficient troops to take over our sector in Holland, our Battalion was re-assigned to the 29th Infantry Division. The 29th Division also was in XIX Corps at that time.

We departed Guttecoven on 6 November and travelled 18 miles to Boutsch, just outside of Heerlen, Holland. The next day we travelled 10 miles to "C" Battery's new gun position at Merkstein, Germany. Infantry companies had their bivouacs all around our gun positions. Merkstein is 8 km. due north of Aachen and on the Siegfried Line (Westwall), although most of the concrete pill boxes and other defenses were just east of Merkstein. The Siegfried line was soon breached between Merkstein and Baesweiler. Flamethrowers convinced many bunker-defending Germans to surrender. There was one concrete bunker where the enemy would not give up. Hence, it had to be buried along with its occupants. Preparations were underway for the crossing of the River Roer and the drive onto Cologne.

The 29th Infantry Division had been a national guard outfit from Maryland, Virginia, and the District of Columbia. They were on the D-Day landings at Normandy. It seemed to me that generally they were close by the 30th Division. Major General Charles Gehrhardt, who commanded the 29th Infantry Division, was reputed to have three divisions: one in the graveyard, one in hospitals, and one in the field. They had suffered many casualties on "D Day." The Division Cemetery at LaCombe, Normandy, contains more than 2,000 white crosses. General Gehrhardt absolutely, and unequivocally, required the wearing of steel helmets with chin-straps buckled, by all personnel at all times. If General Gehrhardt saw a soldier without a helmet, or possibly some other infraction of rules, he simply turned to the nearest officer and said "What have you done about it?" Should the officer respond "that man is not in my company or battalion." a storm would break loose and the unfortunate officer would learn about the responsibility which goes along with rank. Some unfortunate officers were relieved and sent to the rear. It seemed as though General Gehrhardt was always trying to be like General Patton. It wasn't coincidental that they were the only two generals who were continually accompanied by a pet dog.

General Gehrhardt was correct in requiring the wearing of steel helmets, with chin straps buckled, at all times. However, there are better methods of leadership than relieving the nearest lieutenant and sending him in disgrace to the rear echelons. The lieutenant's lifetime was in ruins.

General Gehrhardt required artillery forward observers to be with the three-man point in any advance. Most other divisions left this decision to the observer. When the artillery observer was with the point he frequently was pinned down by German fire, and unable to even raise his head. Someone farther back in the column usually had to rescue the point by adjusting artillery fire on the Germans. There were times when the point could no longer be rescued. Well-trained German soldiers usually were more interested in the larger contingent of troops which was following the point. So the fire exchange could be over the heads of the point. These tactics were briefly mentioned in Chapter 2.

Baesweiler Church tower observation post

We maintained our gun positions at Merkstein from 7 until 20 November. My observation post was in the twin church tower at Baesweiler. The road connecting the two towns became known as the "bowling alley", because it was under sporadic German direct fire from 88 mm guns. Day time travel was hazardous, but if one drove fast enough they could get through. Night travel was not possible because of patrols. More than one American vehicle was hit, but our teams always survived the trip, along with some hair-raising experiences. Most roads became deeply rutted quagmires. The few paved roads were rapidly converted to rutted dirt roads by the destructive action of steel tank treads.

Baesweiler church was a scary place. Germans frequently shelled it. The twin-towered Church was solid brick and comparatively safe but the sounds of bursting shells reverberated and echoed throughout the structure. Some shells penetrated the main roof and the acoustics in the chancel repeatedly amplified the sounds of each burst. I can still hear them fifty years later. In the 1990's I, and many other veterans, require hearing aids.

The terrain was flat as a billiard table. Church towers are not the safest observation post, but we had no alternative. The view from the southern tower included the towns of Bettendorf, Siersdorf, and Setterich. One time, I was forced to leap from the tower to the first landing when German shells were zeroing in on the tower that was somewhat separate from the chancel. I suffered a sprained and swollen ankle but refused hospitalization and hobbled back to the tower two days later.

Our dining and sleeping area in Baesweiler was in the basement of a German home, which as usual had a stone arch ceiling. There were ample stores of preserves to provide us with better food than our "C" rations. The populace had been evacuated deeper into Germany. Rebuilding of the Church was nearing completion at my 1956 visit. Other destruction was still evident. This was a coal mining region. Many towns, including Baesweiler, had immense piles of coal slag towering above the houses. These slag piles were unstable and usually not suitable for observation posts.

The twelve-November-days' drive to the Roer

The 29th Division, with the 2nd Armored Division on its left, commenced the big attack on 16 November despite the inclement weather and oceans of mud. On our left, the 2nd Armored Division's tanks could be seen struggling through the muddy terrain at slow speed, and probably in low gear. Each small farming town was a German strong point. The immense fields of potatoes and sugar beets, which surrounded the towns, allowed for interlocking cross fire by the German defense system. Our advance was slow against strong German resistance in each farming town. Allied air forces dropped thousands of tons of bombs on our objectives. It was reassuring to watch P-47 airplanes, with their tremendous bomb load, devastate towns which were our objectives. However, the Germans took temporary shelter in the stone-arched cellars and then came out fighting. Direct bomb hits would break through the cellar stone arches. Artillery had little effect. The 29th Division infantry suffered heavy casualties, as did the 2nd Armored Division on our left. Major General Ernest Harmon, Commanding the 2nd Armored Division, later wrote:

> The Second Armored's twelve-day drive to the Roer River gets little space in history books although some believe it was the most titanic U.S. tank battle of all time. It certainly was the toughest fight of my military career (Harmon, p. 215).

Ninth Army lost 84 medium and 15 light tanks. Most of these were in the 2nd Armored Division. German forces lost 110 Tiger and Panther tanks (Infantry Journal. 1947. p. 113).

Eventually we had the rubble which had been Bettendorf, and then we captured Siersdorf. Dead American and German soldiers remained where they fell. One American remained propped in a doorway at Siersdorf for several days. Our graves registration people could not get into areas still being fought over. The infantry battalion I was with requested cancellation of the scheduled bombing of Aldenhoven because of our sudden rapid movement. Artillery had already done sufficient shelling of Aldenhoven. The few remaining buildings soon became aid stations and headquarters for our forces.

The Aldenhoven gun position, Hasenfeld Gut, and the Sportzplatz

Battery "C" moved the six miles to Bettendorf on the 20th of November, and a few hours later moved to the isolated farm house northeast of Aldenhoven. Infantry troops were still fighting in Aldenhoven as we moved in. We didn't realize that this was to be our home for two months. The objective of the 29th Division was to cross the 65-foot wide Roer River at Juelich. There were two problems. The Germans controlled the up-stream dams (the river flows north) and could flood out any crossings, thus isolating beachheads. The other problem was that the Germans were deeply intrenched in the Sportsplatz and the farmstead Hasenfeld Gut on the west side of the Roer. The Sportsplatz (athletic stadium) for the town of Juelich was a large concrete structure surrounded by a high concrete wall and a water-filled moat. This was further strengthened by a barren plain on the west side. The designer of the Sportsplatz may well have anticipated its eventual use. He could not have designed it better. Yet, Julich was easterly of the Siegfried Line.

Our Battalion was in direct support of each of the 29th Division's regiments at one time or another. I served as a forward observer for a few more weeks with an observation post at Koslar. It was another miserable farming village which was under continual German artillery and mortar fire. Koslar was a critical launching point for our repeated attacks on the Sportsplatz. Many of Ninth Army's 1,133 KIA's in this operation were lost in attacks on the Sportsplatz. German dead were estimated at more than 6,000 of whom 1,300 were actually buried by our combat forces who could not wait for Graves Registration units. We buried some German soldiers at our gun position when we arrived.

Von Zangen's headquarters at Koenigshofen. Unbeknownst to us at the time

General Gustav von Zangen commanded the German Fifteenth Army. He deployed the 12th SS Corps on the north opposing the British forces. The

LXXXI Corps was deployed opposite the American Ninth Army. The Corps, under command of General Koechling, included the heavily battered 246th Infantry Division, the efficient 3rd Panzer Grenadier Division, the newly arrived 47th Volks Grenadier Division, and the backbone of the Corps—the 12th Volks Grenadier Division (von Zangen 1947). The LXXXI AK not only had massive artillery, but they also had an "amazing supply of ammunition". In addition to the divisional artillery the 766th and the 408th Artillery Corps each had 75 guns. The German heavy artillery was most impressive. Included were three Russian batteries of Fortress Artillery, one Czech 24cm Fortress gun, and three super railroad guns. Two of these were 28cm Bruno railroad guns which had been built in 1937. They had a maximum range of 29.5 km. with their 11.2 meter barrels. The third super gun was a K5 28cm railroad gun, which was the most powerful German-built gun used in the war. Two of these guns were captured by the Americans at Anzio. This mammoth gun weighed in at 218 tons. Each conventional shell weighed 255 kg. The rocket-assisted projectiles, with a range of 86.5 km, weighed 248 kg (Engelmann 1976).

The other Corps in the German 15th Army also had similar artillery support. The German forces opposing the American Ninth Army also included Artillery Observation Battalion 14 and one Fortress Observation Battalion. These units employed flash-sound and other artillery-detection systems to locate each American artillery battery (von Zangen 1947).

A Volks Artillery should not be confused with the German Volksturm. The Volksturm was the last available manpower, including children and the aged. A Volks Artillery Corps, manned by ordinary soldiers, consisted of six battalions, each with three batteries of four guns. They were mixed light and heavy artillery (German Military Studies p.20).

The German 49th and 246th Infantry Divisions were ensconsed in the west edge of Juelich (vonZangen 1947: p.12) and the 10th SS Panzer Division (Schramm 1946:p. 166).

The loss of the "important Aldenhoven" to the American 29th Division, only six days after von Zangen had established his command post at Koenigshoven, disturbed the Germans greatly. Aldenhoven had been defended by the German 246th Division. When von Zangen learned that a light field artillery battalion had gun positions northeast of Aldenhoven he was in a frightful mood. "Thus Juelich is within range of enemy (American) light artillery" (von Zangen 1947: p.16). These batteries would have to be neutralized. The most advanced parts of von Zangen's artillery was in the Lucherberg area (von Zangen 1947:p.17).

Juelich's national title

Once we were in position northeast of Aldenhoven all of the town of Juelich was within 5 km which meant very easy range for our howitzers. This meant we could use lower charges, attain higher trejectories, and have more accurate fire. The town was entirely on the east side of the river except for their sportsplatz. Von Zangen was furious to learn that American light artillery could now engage his two or three divisions which were deeply dug in throughout Juelich. He had already experienced the highly sophisticated procedures which American artillery were using. He also knew that we could change any artillery round from fuze quick to fuze delay with the aid of a screw driver. Our 105mm shells, with fuze delay could penetrate the stone-arch cellar ceilings his troops were relying on for protection. This was particularly true at close ranges when the shell trajectory was high. At longer ranges, where flatter shell trajectories were required, fuze-delay tipped shells were likely to ricochet.

Neither von Zangen nor Colonel McDonald realized that in the very near future Time-Life Books would refer to Juelich as

"THE MOST RUINED TOWN IN GERMANY"

(Botting 1983:p.51)

This title was attained through the 283rd Field Artillery Battalion. When I returned to the scene in 1957 I was greeted with nothing but the harshest of stares from the citizenry who were trying to rebuild. Even the cellars of the homes were 97% destroyed. Much of the town has never been rebuilt. Juelich no longer appears on small-scale highway maps.

The 283rd Field Artillery Battalion command post

Colonel McDonald also realized the importance of our position. He turned to his operations officer and stated "be sure our Battalion history includes that at this position our guns were the most eastern of allied artillery on the Western Front, and the closest allied artillery to Berlin".

Probably, Colonel McDonald was aware of the German artillery observation battalion's capabilities. He knew that the Germans accurately detected our batteries and engaged then with accurate counterfire. For example, "B" Battery had moved their guns into position near Aldenhoven.

As the 29th Division infantry moved slowly eastward, "B" Battery was assigned positions up close to the infantry. We put our four howitzers in a small swale (hollow area) and dug in. On a ridge, about 200 yards to our front, two American Sherman tanks were engaged in a firefight with German forces. For at least 12 hours "B" Battery was heavily shelled by German mortars and artillery. One German shell hit a tree which was next to our kitchen truck where the mess sergeant and his men were working with only canvas protection. John Kopcho, Fate Evans, and Telford Fletcher (mess sergeant), were badly wounded. They were evacuated, and never did return to us. At about the same time Frederick W. Nover, of Roanoke, Virginia, was killed up in the gun position. Five other men were seriously wounded and were evacuated through one of the infantry aid stations. They included Staff Sergeant John Lowry, and Sergeant Norman Davis. Several trucks were damaged, including one Service Battery truck. Sergeant Michael Patrick of Service Battery was seriously wounded while he was delivering supplies to "B" Battery. His supply truck, including some duffle bags therein, were riddled with shrapnel. Captain Sullivan could not reach battalion headquarters because telephone wire was still being installed. Radio contact, as frequently happened, was out. Thus, he decided, on his own volition, to move the battery to a safer location. Shortly after the move the recently evacuated positions were pounded by German artillery. The new position was within range of all fire missions. Several days later we returned to the original position. Other "B" Battery men wounded at about the same time included Peter Kobar, Howard Muller, Harold Levanda and Robert Reeves. The names of the men who were evacuated through aid stations other than ours do not appear in the listing of Purple Hearts awarded in our 1945 history (Denton, 1994).

Battery "C" gun position

Battery "C" was the one which General von Zangen was most concerned about after "B" Battery's move. It was furthest east. The battery received a great deal of attention from German gunners. The other batteries were not overlooked. I became battery executive in early December. This is the artillery officer in charge of the gun position. German shelling was intense. More than 500 shells landed in our gun position, and many more in our kitchen and latrine areas. Some were duds and never exploded. Most incoming shells were preceded by a whine of several seconds duration. Sometimes we could hear the distant gun fire and then time the flight. The exact distance to the enemy gun could only be determined if we saw the flash of their firing and subsequently determined the seconds until we heard the noise of that flash. Then, the speed of sound (corrected for temperature and other factors) could be used to determine the distance from our position to the German battery. One day I was talking with the battalion fire direction center by telephone, when a barrage commenced. I held the talk button of the telephone down and let them listen to all 18 rounds whistle in and then burst. This dramatised the importance of their taking countermeasures such as aerial detection of the enemy gun positions.

Despite the cold wet rain and the deep mud we were superbly dug in and suffered relatively few casualties in light of our being the prime target for von Zangen's immense artillery force. Our foxholes were covered by doors from nearby houses and then mounds of dirt. Once, when we heard the whistle of an incoming round, George Glarner shouted "They won't get me" and jumped into his well-covered hole. The shell hit directly on Glarner's foxhole. Fortunately, our battalion medical detachment had decentralized during this period and assigned one medic to each firing battery. Casper Migliori of Monroe, Michigan, was on hand to render first aid to George Glarner. He was badly wounded, but three months later he had recovered and re-joined us. It was good that so many of our hospitalized soldiers made every effort to return to their own.

One German shell hit in the well dug-in ammunition pit of DeFelice's number two gun position. The explosion split three of our 105 mm howitzer shells, and knocked fuses from two others. Fortunately, our shells did not blow up. Ordnance officers, examining the scene, were amazed that none of our shells detonated. This was luck! Tom Graden was not as lucky when the white phosphorous shells he was unloading took a direct hit. Fortunately, Tom survived. Uncountable German shells were duds (they hit in our area but did not explode).

Our executive hole was deeply dug in with good overhead cover, and even electric lights. We kept our 3/4 ton truck close by because of the radio mounted therein. Flashlight bulbs were connected to our six-volt vehicular battery, which caused little drain on the battery. The flashlight bulbs glowed brilliantly. Each of us had a ledge for sleeping. We had telephonic communication to each howitzer, to our battery command post and to battalion fire direction center. This was all backed up by radio communications.

Usually, I slept so soundly I didn't hear night-time incoming shells. The men marvelled at that, and the next morning would show me the craters. There were many night-time fire missions. Howitzers were manned at all times.

General Sands, the division artillery officer, frequently visited our position. He was highly impressed by our sandbagged fortifications. Major Fred Dieterle, and other members of our battalion staff, regularly visited us while we were at Aldenhoven. Our battalion staff officers, following the example of our battalion commander, carried M1 rifles, and regularly toured howitzer positions. Other batteries had more casualties than we.

The listening post on the Roer

Personnel from our battalion, along with an Infantry scout patrol, manned a listening post in the remnants of a German farm house on the banks of the Roer River. It could only be accesssed during hours of darkness, and even then there were German patrols to contend with. Our observers locked themselves into the structure because on several occasions German patrols attempted to search the building and there would be a rattling of doors. Lieutenant Grohne relinquished the post to Lieutenant Gerry Gauthier on Christmas Eve. German patrols were especially obnoxious that night and Gerry had to call down our own artillery fire on his post to drive the Germans away. It wasn't the pleasantest location to be on Christmas Eve.

German Paratroopers and Aircraft sent to wipe us out

German paratroops were a part of the Luftwaffe (AirForce), and not the Wehrmacht (Army), as in our forces. However, von Zangen somehow arranged for paratroops to attempt to wipe out that artillery which so bothered him. It was New Year's eve. There was considerable aerial activity. One Junkers JU-52 trimotor plane, loaded with 14 German paratroopers, attempted to come in low over our position and drop on our rear area. We were fully alert and our machine gunners downed the plane 200 yards in front of our position. The Germans were all killed on impact.

Our 50 Cal. machine guns were deployed as anti-aircraft. Overhead German planes caused a hail of bullets to fly into the air from all parts of the Division sector. Naturally what goes up also comes down. Sergeant Stratton was hit on his head (he was wearing a helmet) with one spent shell, but other than a black face he was all right. The Germans launched numerous aerial sorties on New Year's Day. One bomb fell close by a large all-glass office building in Heerlen. Most every window pane was shattered. It was one awful mess. Twenty-seven German airplanes were shot down in the 29th Division's sector that New Years Day. Stanley Wardach and Joseph Morrelly brought down one tree-top flying German plane at mid-morning (Wardach 1993). They also participated in bringing down the paratroopers. This was a great start for the year 1945.

Possibly some of our Division-directed unobserved-fire targets were the results of American sound-flash artillery observation reports. However, neither I, nor the Battalion Survey Officer, Lieutenant Philip Lull, knew of any American sound-flash work to allow us the counter-battery capability which the German forces had (Lull 1994). Apparently German artillery observation battalions communicated the location of American gun positions directly to the artillery battalions whereas our artillery observation battalions reported to corps artillery headquarters. Communicating the information directly to those who are under fire usually obtains the quickest action to remedy matters.

The Sportsplatz and Hasenfeld Gut

It seemed to me that General Gehrhardt erred in repeatedly sending in only one battalion, and then relieving the battalion commander after the mission failed. There were many needless American casualties. Alternatively, these casualties were more than enough to have blockaded the Germans and then simply by-passed them. Certainly the Germans would have withdrawn these forces before they released the waters from the upstream dams. Most infantry officers, with whom I spoke, felt that Gehrhardt should have used more than one battalion at a time. But, in any case we could not cross the Roer until we had control of the upstream dams. The whole mission seemed futile in consideration of the German-controlled upstream dams.

Repeated attacks on the Sportsplatz and Hasenfeld Gut during early December failed to dislodge the Germans. Casualties on both sides were very high, especially in the 3rd Battalion of the 115th Infantry Regiment. Both objectives were taken after six tries.

In 1957, I found the site had been designated as a memorial. Germans, visiting the site, where their loved ones had departed this earth, were drinking beer, and even wine, to a background of soft classical music at a gasthaus on the west bank of the Roer adjacent to the crumbled Sportsplatz. The only trees were newly planted saplings. Swans glided along the placid river. I, driving up in my 1951 convertible Rambler, drew numerous cold stares. The waiters were extremely polite, however. They may well have been displaced persons.

Return to the German Koenigshoven Command Post and unbeknownst to us

Von Zangen had the dual mission of relieving the 5th Panzer Army and also preparing to launch an attack two days after the Ardennes Offensive. He would need as much territory as possible on the west bank of the Roer to launch the offensive. The Germans still held the Heinsberg salient opposite the British forces who had replaced the 125th Cavalry in Holland. The German plan called for the 9th Panzer Division and the 176th Infantry Division to assault from Havert (two km northeast from Millen Castle) passing Sittard to the west, reaching Maastricht, and capturing the bridges over the Maas River (von Zangen, 1947: p.41). The Juelich bridgeheads west of the Roehr (the Hasenfeld farmstead and the Sportsplatz) were also vital to launching the planned German offensive.

The Germans withheld revealing plans for the Ardennes Offensive to Corps and lower levels until December 10th. Eventually, it became known that a major offensive was imminent. The German troops had lived through the trials and disappointments of the Allied invasion, the disgraceful period of flight, and the surrender of their native soil. They sensed, however, instinctively that even if this plan would succeed there was no chance of winning the war. They were prepared to give all to bring the disgraceful period to an end. They wanted to regain their own country's boundaries and thus fight to obtain as soon as possible bearable peace terms for their people. Von Zangen knew his troops had full confidence in him (von Zangen 1947: p.43).

The German Supreme Command in the wet and gloomy Gorlitz forest (Rastenburg, East Prussia)

Hitler and his generals were in accord that a systematic examination of all the possibilities for an offensive pointed to the Western Front (Schramm 1946: p. 51). This led to their Ardennes Offensive.

> In the event that the plan outlined on paper could be entirely transposed into reality, the entire British Army Group as well as the American units fighting in the vicinity of Aachen, Juelich, and Dueren would be cut off from the rest of the Allied forces (Schramm 1946: p.80)

On 10 November a new army group — Army Group H was activated under General Student. It included Fifteenth Army and the German First Army on the lower Rhine. This Army Group was to carry out the secondary thrust of the Ardennes Offensive. This was referred to as the "Little plan". The annihilation of American forces in the triangle of Sittard— Liege— Monschau (including our 29th Division) was intended (Schramm 1946: P.115).

On 3 November General Model was more firmly convinced than ever that "little" plan was better than the one elaborated by the Fuehrer's headquarters for the Ardennes Offensive. Fortunately for our 283rd Field Artillery, Hitler insisted that the "little plan" would be secondary to the Ardennes offensive, and even then it would only be launched if the Ardennes Offensive attained its objectives.

The Ardennes offensive

The Battle of the Bulge (Ardennes Offensive) in mid-December was immediately to our south. There was German pressure in our sector, but we maintained our position and thus held the north flank of the bulge. Our lines were extended to permit neighboring units to join the battle to our south. The 29th Division went on the defensive and were re-assigned from the XIX Corps to the XIII Corps. Ninth Army was placed under the control of the the 21st Army Group under Montgomery.

About 20 December, when it was realized that the Ardennes offensive had failed, German Army Group B declared the attack from the Heinsberg salient ("the little plan") could no longer be executed (von Zangen 1947: p.49).

German rockets

Several times we observed V-1 rockets heading towards Amsterdam or Liege. They were low flying, noisy, and relatively slow. Our fighter planes were frequently able to shoot them down. The V-II rockets, that went into outer space were far too high, and also too fast, to be shot down.

"C" Battery command post

Our battery command post (CP) was in the stone-arched basement of the adjacent farm house. The battery clerk, Zane Hoffman of Harrisville, Pennsylvania, was a good piano player. The piano was moved from the demolished upper floor into the basement, and our CP enjoyed music. After dinner, I frequently was in the battery command post censoring the day's outgoing mail. Every letter had to be examined to prevent troops from stating where we were, or any details of our travel.

Incoming mail arrived regularly as long as we were in a stagnant position. It was difficult for the home front to decide what to send. Eventually, they learned that dry wool socks were better than cookies and candy. Parcels were received from families as well as a wide variety of family friends, churches, and school teachers. Naturally, parcels were shared.

Our kitchen was set up in the stone farm sheds. Meals were good, but frequently interrupted by German shelling. Our one Texan brought in a cow and showed us how to butcher it for fresh beef. Carrying of rifles and wearing steel helmets at all times was enforced.

One perspective on the 29th Division

Stanley Frank wrote a lengthy article on the 29th Division in the Saturday Evening Post of March 16, 1946. He cited Major General Charles Hunter Gerhardt as being a showman with General Patton characteristics, Frank wrote of the heavy casualties incurred by the Division in spearheading the Normandy invasion at Omaha Beach. Overall they caught more artillery fire than any other American Division in Europe. Total casualties for the entire war were 20,688.

"Gehrhardt was a fanatic on discipline"

"Gehrhardt busted officers indiscriminately and threw the book at enlisted men who neglected to shave or keep their vehicles and gear clean."

"At Gehrhardt's insistence, officers wore the insignia of their rank on field jackets and helmets in combat. He conceded that such markings offered splendid targets for snipers..."

In the drive for the Roer River in November 1944, Stanley Frank's quotes include:

"Resistance was unexpectedly tough across the flat terrain, studded with a string of villages so close together that each knot of houses was a strong point offering supporting cross fire against an attacking force."

"The German fanatic defence of the Roer River was related to their buildup for the Ardennes offensive."

"On Christmas night, 1944, KP's crawled across the snowy beet-fields with turkey, peas, and mashed potatoes for men on guard at the river."

"To take Julich, the 29th had first to take its Sportsplatz — and after, six tries took it in a night attack."

A rusty machine gun barrel and leadership

I don't recommend the Patton/Gehrhardt style of leadership. It does permeate through the ranks, and not always for the best. As an example, while rummaging through one of the ammunition trucks at Aldenhoven I found a spare 50 caliber machine gun barrel. These were carried to exchange with barrels which were getting too hot from firing. Asbestos gloves were used to effect the exchange. This one barrel was about as rusty as they can get. The thought which immediately struck me was, what would have happened if General Gehrhardt, or one of his staff, had found this dereliction of duty? I would have missed out on the making of history with a first-class bunch of guys. I immediately relieved the ammunition section corporal. I never stopped to think that he probably felt the same way and, that although he would continue with the same guys, he would not have held a leadership role in the Great War. Yes, I supppose my action was motivation for all crews to delve into the depths of their trucks and see what maintenance was needed. But later I realized the matter could have been settled with better leadership on my part.

Martin Williamson, my battery commander for the entire war, was an example of good leadership. He was not bossy, but tactfully recommended corrections when they were needed. Most important was, that he allowed me to use initiative and ingenuity. He backed me up in the machine-gun barrel affair. And again, his style of leadership, although in conflict with Gehrhardt's, permeated the ranks. As long as the gunnery sergeants were continually improving their fortifications at Aldenhoven I didn't tell them what to do. I did offer suggestions, including "walk over and see what the 1st Gun Section did today".

Never once during the war did I think about my efficiency reports. No one ever discussed the matter with me either. Never once did I receive the slightest criticism nor even suggestion for better performance. And, I never heard the matter of efficiency reports mentioned by anyone. In later years I learned that Captain Williamson had been most generous in whatever rating report he used for my performance. Colonel McDonald, the endorsing officer, also was most generous. When I applied for the Regular Army in 1949, I was readily accepted. To this day, I don't know what the system was nor how well other officers fared.

Every person is an individual. We are all as different as Bentley's snow flakes. Some individuals need firm discipline, but our Army doesn't need very many soldiers of this type. Individualists can become the best of soldiers, if they are properly led. The Gehrhardt style doesn't recognize the differences in people. It assumes they all can be motivated by scare tactics. Germanic-Prussian-Teutonic firm-discipline-leadership forged many victories for the Germans between 1939 and 1944. However, Germans are not Americans. The Frederick Jackson Turner Thesis of 1921 focused on why Americans are a unique people.

Von Zangen's problems in the wake of their failed offensive

In the middle of January von Zangen was confronted by the need to assemble and deploy reserve companies.

> These choice elements were naturally doomed by detaching them from their familiar battle unit. He ordered collection of all men returning from leave (and medical leave) in especially selected collection points without regard as to their original troop organization, requirements of one's own troops or branch of service. Employment at focal points. Again failure. Court martial measures against those, who "escape" to their own units, makes matters worse (von Zangen 1947: p. 31).

Adieu to the 29th Division

The 29th Division, now on the defensive, had little need for Theatre Artillery. On the 26th of January, 1945, our 283rd Field Artillery Battalion was re-assigned within the Theatre. Probably the longest move of any artillery battalion from one gun position to another in a combat situation followed.

At this stage, the Campaign for the Rhineland was still not history. It will be continued in Chapters 5 and 6. The next chapter is a hiatus for an intermediate campaign.

In my 1957 revisit to Aldenhoven I found all foxholes had been filled in, the farm house had not been repaired, and numerous refugees were living in the adjoining farm buildings including barns. Housing was still critically short in the British Zone of Occupation. My visit did cause a flurry of excitement. I did take photographs which I was unable to take during the war.

Chapter 4
CAMPAIGN FOR ALSACE-LORRAINE

Germany west of the Rhine River

A great deal of Germany — even today — is west of the Rhine River. In January 1945, German forces continued to hold much of this terrain. This included the German Saar along with a strip of France to the south of the Saar from Sarreguemines to Hagenau. German forces were trying to recapture Strasbourg, Capital of Alsace. Strasbourg is only 30 miles south of the Saar. We'll return to the Saar in Chapter six.

In January 1945, the German forces also held a sizeable beachhead on the west side of the Rhine around Colmar. Colmar is about 40 miles south of Strasbourg. This Colmar pocket of Germans was a potent threat to any Allied Rhine River crossing in that sector. SHAEF decided that this pocket had to be eliminated.

For centuries Alsace-Lorraine had been hotly contested by France and Germany. France was forced to cede it to Germany in June 1940 when Germany dictated the peace terms to the Vichy French. Some of the Alsace-Lorraine populace were supportive of Germany and welcomed German troops. Others were not. No election was held in 1940 and it is not possible to state where most stood on the matter. The 1940 action might better be termed retro-ceded, because Germany was forced to cede it to France after World War I. It again became a part of France after the Second World War.

Teenagers of Alsace-Lorraine in the SS

Young men of Alsace-Lorraine were conscripted into the German Army after June 1940. This was contrary to the 1940 Armistice terms. Some evaded conscription by escaping to Switzerland or 'unoccupied' southern France. Some volunteered for the German Army. Those born in 1926 were called up in January 1944 and many ended up in the SS Das Reich Division. These Alsatians dominated the Das Reich Division force which hung 99 innocent French men and teenagers from lamposts in the French town of Tulle three days after the Americans invaded Normandy. And, the next day committed the worst atrocity of the War on the western front at Oradour. Both incidents were Satanically ritualistic. In the Tulle affair the young men were assembled in the school yard where the priest administered the last rights of the Church to each of them. They were then marched in formation through town stopping at each lampost or balcony to hang one more victim. The Oradour affair was even more sickening. The village population of 642 men, women, and children were brutally burned alive in the town church or killed. Every building was burned. Today, the village remains as it was after the atrocity. One elderly woman made her escape from the burning Church. Madame Rouffanche still survived in 1979. There were a few other survivors. It was one more sickening story of man's inhumanity to man. The decision to proceed with this operation was made by a low-ranking German officer who had recently lost a friend to the French Resistance. The January 1953 trial of the culprits in Bordeaux was about as bizarre as court trials can get. None of the accused was executed. Very few were punished. Were they innocent Alsatians or Satanic Germans?

Himmler needed young men for his Waffen SS, and so long as requirements for the Wehrmacht were paramount inside Germany, he was forced to turn elsewhere (Stein 1966;p. 144). The SS divisions which were organized in 1943 included lads as young as 16. Two SS divisions boasted an average age of 18 including officers (Stein 1966;p. 207).

Overall, most of the Alsatians conscripted into the German army were dispatched to the eastern front. This discouraged them from breaking their irrevocable oath of loyalty to Adolph Schicklegruber. Many who were not killed were taken prisoner by the Soviets. Few ever returned to Alsace-Lorraine.

The Free French, under Charles DeGaulle, considered the concession of Alsace-Lorraine to have been illegal. Alsatians were still considered French as far as they were concerned. French General LeClerc, and possibly others, considered captured soldiers from Alsace-Lorraine to have been traitors. Some of them, taken prisoner, were immediately shot, without trial, on orders of General LeClerc (Whiting, 11th photo page following page 86).

Sixth Army Group, First French Army, XXI Corps and 28th Infantry Division

The U.S. Sixth Army Group was commanded by General Jacob Devers. It was on the right flank of the Allied Armies as they advanced through Europe. It included the U.S. 7th Army, and the French 1st Army. The First French Army was commanded by General d'Armee de Lattre de Tassigny. After the downfall of France in 1940 he had remained in the small French force under the Vichy Government. In 1942 he had been imprisoned for opposition to the German occupation of southern France. He escaped and joined DeGaulle in England.

One of the corps assigned to the French 1st Army was the XXI Corps commanded by Major General F. W. Milburn. Bear in mind, that a corps is a headquarters unit to which varying divisions are assigned from time to time.

One of the divisions assigned to XXI Corps, at this time, was the U.S. 28th Infantry Division. This had been Pennsylvania's National Guard Division. The former Pennsylvania National Guard also included units other than the 28th Division, such as the cavalry regiment with which I enlisted in June 1939. The 28th Division landed in Normandy in late July and had fought through the hedgerows where their commanding general had been killed. Now, Major General Norman D. Cota commanded the Division. The Division had recently been badly mauled in the initial German strike of the Ardennes offensive. Since then they had received many replacements and now were ready for their own offensive. Their three regiments, which we supported with artillery fire were the 109th, 110th, and 112th. Note that as usual, we were assigned to a division and had very little to do with any corps headquarters.

Our movement to the Battle of Colmar

Our 283rd Field Artillery Battalion was assigned to support the 28th Division in this attack. We left our Aldenhoven gun positions on 26 January and made the longest combat move in history from one gun position to another. The Witt Diary does not include the many towns we passed through other than Verdun and St. Die (Map #5). Our new gun positions were to be at Ostheim, seven miles north of Colmar. Travel was hampered by snow storms as far as St. Die. The weather was frigid and we all had a pretty miserable trip, doing whatever we could to keep from freezing in our unheated vehicles. Then matters became even worse. The 35-mile trip over the Vosges mountain pass, with narrow, steep, winding roads, and numerous hairpin curves from St. Die to Ostheim was complicated by a blizzard of record proportions. Senior officers of our battalion had gone on ahead to coordinate our mission and locate gun positions. Battery "C" led the main convoy over the mountain pass. Thus, responsibility for getting the convoy through the blizzard fell on my shoulders. At most any other time the scene would have been picturesqe. Dense stands of amazingly tall, and straight, silver fir were clothed in fresh white snow. French roadworkers, and local people, were superb in plowing a path for our movement. Mechanized snow removal equipment was not available. Shovels and hand held snow pushers were used. We moved slowly, even manhandling some trailers and howitzers around hairpin curves, and eventually made our way down into our Ostheim gun positions. There was no complaining from our troops. It was amazing to see how supportive they were of this young lieutenant. Their suggestions and encouragement were offered in a manner which reflected highly on their upbringing.

Our Service Battery, with its ammunition train (not really a train, but a number of trucks used only for ammunition supply) was usually to the rear of Battalion Headquarters. However, they suffered as many casualties as some of the firing batteries. For example, they had to retrace this route each time they went for more ammunition because the main ammunition dump was in St. Die.

> I was driving the lead truck delivering ammunition to a firing battery in the Battle of Colmar. January 29, 1945 was very cold. Fred Kimmich was sitting in the middle, between Les Snyder and myself. We came under an artillery attack near Reipertswiller. One air burst exploded very near our truck. Shrapnel went clear through Fred and then struck me in the lower right side. Fred was killed instantly. Les Snyder was also wounded by shrapnel (Lacy 1994).

> The night of the day Fred was killed we were going back over the mountains for more ammunition. The road was icy and one uncontrolled slide meant the truck would be over the edge and fall into the valley below. At the sharp hair-pin turns in the road there was always a heavy truck anchored on the inside of the turn. Vehicles rounding the turn were first connected to the winch of the anchored heavy truck to secure them as they made the turn. A driver of one of the anchored trucks saw our 283rd Field Artillery logo and asked if Fred Kimmich was there — "I am his brother". It was hard to tell him that Fred had just been killed (Parker 1994).

> From the foot of the Vosges to the Rhine River is a flat plain. The Germans could see us in the daylight so it was best for our ammunition train to travel at night with only blackout lights (Parker 1994).

Service Battery also had a number of wounded when they were shelled in the postion they had just left up north. And, to top all that off, the ammunition train was shelled by German railway guns a few weeks later when they departed Colmar. Some of these incidents may have been German unobserved fire, where they had zeroed in certain road junctions and then periodically fired interdiction fire. Other incidents may well have been German observed fire.

Unheated vehicles

Some readers may wonder about our unheated vehicles in the war. First, most vehicles were designed for easy, and rapid, exit as well as for shooting weapons from within the vehicle when necessary. They had canvas tops, and even canvas side curtains, for very inclement weather. Heat would have dissipated rapidly in such conditions. Most vehicles did not have windshield defrosters. Perhaps this was also linked to civilian vehicles of the early 1930's. Automatic defrosters and heaters had been available in some models for several years. They were added to all automobiles and trucks starting in the late 1930's.

Our gun position in the Battle of Colmar

Ostheim was a small town, seven miles north of Colmar. It appeared we were in the enemy terrain, because our arrival was not to the tune of church bells. There were no civilians at all. We might just as well have been in Germany, but then many natives considered Alsace-Lorraine Germany. However, there were no white sheets hanging from the homes as we had found in Germany when we were with the 29th Division. The 2nd of February 1945, Battery "C" moved forward to a gun position at the north edge of Colmar, just west of the main highway. The ground was covered with five inches of fresh snow, which was but a fraction of the snowfall in the mountain pass we had recently come through. Our battery CP and kitchen were set up in a vacant farm house 200 yards in front of our howitzers. This was necessary in order to provide shelter for our kitchen and still have proper fields of howitzer fire. There was virtually no danger to our kitchen crew from any of our firing, although the noise was bothersome. We never had any rounds explode early. American defense workers were reliable people and thus our shells were never defective. In between fire missions we dug in and covered our holes with doors, boards, and other material from vacated French homes. Then, we piled dirt on top of all that. The ground was not frozen except for the first few inches. Alsace apparently has a milder climate than most of France. We always assumed we would be in a position for an indefinite period of time, and continued to improve our protection.

By the 3rd of February, a pleasant warm spell had the snow melting. We found the path from our gun position to the kitchen was mined with German Teller mines. These are anti-tank mines and usually don't detonate from the weight of people. Nevertheless we found another path.

Our soldiers and their equipment

Footwear was a continual problem during the war. In 1941, the Army changed the color of boots from brown to black, primarily so the Navy, Air Corps, and the Army could use the same product. Army troops had to dye their boots black. Naturally, black boots absorbed more heat from the sun and in southland training, sweaty feet became problematic. Then for unknown reasons, new black boots were made with the roughened leather, which is usually interior in footwear, on the outside. This ruined any water resistance the boots might have had. It was difficult to understand how any shoe designer could be so stupid. Later I was informed the intent was to prevent soldier exertion in shining boots. Roughened leather can't be polished. Some units required their men to try.

In November 1944 it became apparent that our new boots, in conjunction with the cold wet weather, were causing disabling fungus problems, called trench foot. Rubber overshoes, and shoe pacs, were on order but didn't arrive until late January. In the meanwhile, trench foot caused many casualties. Note that "casualties" are soldiers who had to be evacuated from their unit, temporarily or permanently. Casualties include KIA's (killed in action).

The Army commenced issuing dry socks with the rations, but dry socks didn't help because boots were wet inside. The problem was widespread in Infantry units. Our firing battery, and most likely our battalion, never had any official casualties from trench foot. However, our troops were never satisfied with issue footwear. It seems to this writer that no progress has been made in the past 50 years in the design of military footwear except for the recent design of desert boots for operation Desert Storm. These boots would have been worthless in Europe. In the meanwhile, American shoe manufacturers have made progress in providing waterproof, breathable, boots for civilians hiking the Appalachian Trail or camping in wilderness areas.

An aside note: When I was in Germany in 1954, I contacted a good shoe manufacturer in Chippewa Falls. They mailed me an excellent pair of fleece-lined waterproof leather boots. I still use them occasionally in the 1990's. I expect to be wearing them when I am buried in Vermont.

The Battle of Colmar was a particular difficult time because of the snow, then melting snow, rain, and mud. Anthony Morrelly, of Ambridge, Pennsylvania, had a cold since the long wintry journey from Aldenhoven. Suddenly, his fever soared and we rushed him to the General Hospital in Epinal. He died the next day of pneumonia and was buried in Epinal along with thousands of other Americans. Tony always had a cheerful countenance. He was not robust, but he could usually hold his own on distance marches and other physical challenges. Otherwise, our battalion was a remarkably healthy crew during the winter of 1944-1945. Possibly, the continual living out-of-doors was healthier than the usual winter environment.

American World War II soldiers were significantly smaller, and lighter in weight, than young men in the 1990's. Their overall health would have been rated poor by today's standards. Growing up during the depression era meant adversity, along with a lack of medical and dental care. Many youth had never been to a doctor nor a dentist. One Army dentist at Camp Shelby told me he was averaging seven fillings per hour. The number of men the Army rejected for medical reasons approached one out of every two examined. Possibly some were rejected for eyesight because in "C" Battery only one soldier wore eye glasses. The average WWII soldier was five feet, eight inches tall, the same as in the Civil War. He weighed 148 pounds. These were the men who were not rejected. Variance in height and weight were small. At the time of departure from the New York port our battery had five men who were about six foot tall. Strangely, all five were veterans of the CCC. Could it have been that an adequate diet, such as that in the CCC, was all that was needed for our youth during the depression era? The other men, and all of the officers, were smaller than six feet. There were no muscular men by 1990's standards. We had no problem with overweights. They did not exist. Man-handling howitzers, along with thousands of ammunition crates (howitzer ammunition arrived at gun positions in original wooden crates), was back-breaking work. Every position was continually dug in deeper until the close-station-march order (CSMO) was received. Yet, we never had any knee nor back problems. Our intensive training had toughened the men. We did have at least one dropout back at Camp Rucker.

Greenman was a Philadelphia Jew, small in stature and not as strong as his associates. Yet, he was determined to remain with the battery, hopefully to eventually do battle with the Germans. He was well-liked by all because of his determination. Yet, he would be in a virtual trance after five miles of a 25-mile hike. It really hurt to let him go, but it would have been wrong to have taken him into combat.

We never had any suicides, attempted suicides, homicides, nor self-inflicted wounds within the battalion. I never heard the word "homosexual" and if I had I wouldn't have understood the meaning. Army entrance medical examinations did not query men on their sexual preference as far as I observed. If we did have any men with even the slightest homosexual tendency they never let that be known.

The use of illicit drugs was unheard of. Alcoholic beverages became plentiful as we advanced into Germany. It was not possible to stop the men from drinking and we didn't try. Most of our vehicles eventually had a supply of good liquor, varying from kegs of the finest Moselle down to bottles of schnapps. German military supplies were fair game for confiscation and not considered looting. Yet, I had only one case involving alcoholism. One sergeant had to be reduced in rank and transferred to another battery for drunkeness. His gun crew continually went to extremes to cover up his drunkeness. One might have thought that preference for creating a promotion opportunity might have prevailed over loyalty. This was not so, in this Battalion! Eventually, I realized the situation but cooperated with the gun crew's coverup until the situation became untenable. This reduction in rank was not a court martial, but an administrative procedure. The men realized they could jeopardize their associates by heavy drinking, and by this time, their associates were their real family. Hence, the men were temporate in their drinking with this one exception. There may have been one officer, not in "C" battery, who was given to excess drinking.

Combat with the Germans for Colmar

The First French Army ordered us not to fire at any buildings within the historic city of Colmar. It was hard to comply when the Germans were in buildings. We did our best. Two days later, when the Germans had been forced out of Colmar, they inundated Colmar with artillery shells. This artillery and mortar shelling of recently evacuated sites was a continual pattern by the Germans. Most Allied forces soon learned to expect it.

We saw our first German jet airplane on the 4th of February. Later we learned it was an ME 262. This was one of the first jet aircraft the world had ever seen, and yet our troops readily recognized it. The plane circled above us and we then realized we were the target. Our .50 caliber machine guns couldn't seem to catch up with the circling plane. The pilot had ample time to line up his target. No Allied planes were in the area, and even if they were they could not match the speed of the jet.

We all dove into our shelters and I called the aid station and alerted Dr. Frank Cohen. He asked me to stay on the line as he had his medics start mixing up plasma. I held the talk button down on the telephone and got as low as I could. The whine of the falling bomb was audible for what seemed an eternity. It terminated with a crash, and I and many others realized we had survived. The 500 pounder lit directly on top of our aiming circle in the center of our position. Craters from 500 pounders are large enough to put a 3/4-ton truck in. None of our troops were seriously hurt, but several infantrymen, who were in the back of a truck, 100 yards away, were mortally wounded. Dr. Cohen was at our position in a flash and did what he could to save them. A few minutes after the blast I realized that I was covered with blood. Frank Cohen wanted to treat me first but I assured him it was minor and to devote his effort to the wounded Infantrymen. A bomb fragment had grazed my helmet and encountered an iron stove. I received numerous fragments from the encounter. This was one more example of the determination of our men. They had dug in the howitzers, and their personal shelters, entirely of their own volition. Their efforts had once more been rewarded.

The Germans produced 1600 of the ME262 Jets, but only about 600 were ever airborne. (Seaton 1981: p. 129). This was midwinter and we lived as comfortably as possible. The men were able to scrounge stoves when they were needed. If it was a particularly good stove we carried it along with us. I never had to send the men out looking for stoves nor doors to cover our holes. They did this of their own volition. Sometimes we burned coal and sometimes wood.

The advance on Colmar was proceeding well. The next day, 5 February, our battery took up new gun positions on the east edge of Colmar, adjacent to a neighborhood of prosperous homes. The substantial three-story homes, with stone-arch cellars of course, had been vacated. Our battery CP was lodged in one home, and I and the executive staff in an adjacent home. Our howitzers were in a field next to our house.

Shelling from a captured French 37cm railroad gun

At 11 p.m. there was a large blast which had not been preceded by the customary whistle or whine. I investigated and found that although no one was injured there was a crater close by one of our howitzers. It was large enough to put a 2 1/2 ton truck in. We considered all possibilities, including that of a remotely detonated land mine. A short while later our battery CP house disintegrated. We found Lieutenant Herman Wolf, and his crew, sleeping safely in a demolished first-floor sunporch. They were unaware of the shelling. The Battery Commander and others had been in the cellar and also were safe. It seemed as though we had a German artillery observer in our midst because the battalion headquarters was next on the list.

Bill Davis, with the battalion personnel section, relates the situation as he saw it:

> When we moved into the four-story castle-like former Gestapo Headquarters in Colmar I intuitively sensed we were inviting trouble. Late that night, there were five incoming rounds. The

first round landed in the street in front of the command post (CP). The second hit the roof and blew out the front section of the room where four men were sleeping. The third round hit in the court yard. And, the other two rounds landed in the area of "B" Battery. Major Dieterle and I checked each room for casualties. The four men in the front bedroom were in their sleeping bags which were covered with debris. They were killed by concussion. Colonel McDonald had been sleeping on the third floor. The detonator head was found in his bed. Ordnance reported it was from a 370 mm railroad gun of WWI vintage (Davis 1993).

The Germans had captured a number of French railroad guns in 1940. A few were the 37 cm (370 mm). The 1.6 meter-long shells could be reloaded fairly rapidly by a well-trained crew. Doubtlessly the railroad gun was on a siding of the main-line Karlsruhe-Basel railroad. Within 48 hours this railroad line was devastated by our 105mm howitzers. The Germans had a total of 33 railroad guns in June 1941. Today, the lone survivor is at Aberdeen Proving Ground in Maryland (Engelmann 1976).

Four of us from Service Battery were with Headquarters Battery that night. I picked out a room for us, but before we could bed down Colonel McDonald told us to move because he needed the room for his headquarters men. We moved diagonally across the street to a three-story building which had some nuns and children on the first floor. We bedded down in the top floor dormitory. When we heard the explosion we ran down stairs, past the screaming children, and out to the street. Then we realized the room we had been chased out of was the one which had been hit (Parker 1994).

(Ed. comment: Sergeant James Parker was in effect an assistant to Captain van Valkenburg, Service Battery Commander. He was regularly entrusted with far greater responsibilities than normally would be the case. This is evidenced by his having been the only one of our men who received the French Croix de Guerre award during the entire war. It is difficult to explain why it is that some men shun the usual military advice "to never volunteer" and are quite willing and capable of assuming responsibility.

Lieutenant William Grohne of "A" Battery had been temporarily assigned to Fire Direction Center. He was on the first shift of fire direction center along with the men who were killed. However, he had the good fortune to have just left the devastated area. Fire Direction Center was safely located in the typical basement with a stone-arched ceiling. It was possible to survive a shelling from the largest artillery providing the fuze of the shell was the usual type which causes detonation upon initial impact. Artillery shells can also be equipped with delay fuzes which may not detonate until they have penetrated into the basement. If delay fuzes strike a solid surface they frequently ricochet and detonate some distance away.

German forces typically zeroed in on command posts which they had just retreated from. This probably accounts for the shelling of our battalion command post. The Germans also used sound/flash, survey, and other measures to counter our artillery fire. Probably they had our firing battery's exact position as soon as we fired our first rounds.

I knew most all the men in the entire battalion. We were like one large family after all we had been through. The men killed in Colmar were our finest, most intelligent soldiers. Al Bell from Ozone Park, NY, Sidney Geller of Philadelphia, Sergeant John Muench of Louisville, and Nick Pocetti of Vandegrift, Pennsylvania would probably have gone on to college, and then become leaders in whatever field they chose. Probably all four were making monthly payments into National Service Life Insurance, a governmental program under the Veterans Administration. If so, each of their families received $10,000. along with a Purple Heart medal. Higher amounts were not available. Commercial insurance coverage excluded battle deaths. The $10,000 was the same amount paid to survivors in the First World War. Dependents such as wives who did not re-marry, and children until age 21, received a small monthly pension.

They had all volunteered when their country called them to fight a nation which was bent on world domination, marching to "Deutschland Uber Alles." No, this wasn't Hitler's war. It was a continuation of Bismark's and the Kaiser's program for "Lebensraum" according to Professor Fritz Fischer, of Hamburg University (Fischer 1986). Yes, there were dissenters within Germany but they comprised less than one percent of the population. Thirty five thousand of them were executed, ranging from "The White Rose Group" down to peasants who had reputedly listened to the British Broadcasting Company (BBC) on their hidden radios. Or, even worse, peasants who knew their neighbors were listening to BBC and failed to report them. Yes, even the latter was punishable by immediate death in Judge Roland Freisler's court. Our Air Corps terminated his judicial career. His fellow judges have never been tried for their hideous crimes.

Victory at Colmar

The Colmar pocket of German troops was eliminated by 6 February 1945. Battery "C" took up new gun positions in Niederbergheim, southeast of Colmar, and was assigned targets on the other side of the Rhine River. The French held a victory parade in Colmar and the residents started to return. I could not imagine where all these Colmar residents had been hiding during the battle. In this case we did not see any retribution against those who had collaborated with the Germans. It also seemed that these French had been comparitively well fed.

We moved to Dessenheim, on February 7th, so as to increase our range into Germany. For the next four days we shelled the withdrawing Germans on the east side of the Rhine River, and disrupted rail traffic from Karlsruhe to Basel.

One more payday

Soldier pay was very small, even by the standards of the time. I believe it was about $50 per month for most of the privates and $150 per month for Lieutenants. The exceptionally low pay was a matter of shifting the tax burden to a different generation. Even enrollees in the Civilian Conservation Corps (CCC) of the 1930's had been paid more than military recruits at that time. Yet, I never heard any complaints about the low pay. Motivation overwhelmed materialism!

Most of us had made out allotments to our families and only received trivial amounts in Europe. The currency varied from French francs to Occupation marks in Germany. Gambling was never a problem, except on shipboard. Thievery was not existant. Payday was at the convenience of the tactical situation and not always on the last of the month. We were never paid in American dollars, nor were they available.

Another of our assignments was completed. We were alerted to support the 104th Division (Timber Wolves) under General Terry Allen back north in First Army's sector. Their objective was Cologne.

The campaign for Alsace-Lorraine, along with the Ardennes Offensive, was considered a separate campaign. Eventually this became the second star on our European Theatre ribbon. Unfortunately, an error was made in the official records of the Battalion and the Ardennes-Alsace campaign is not listed among the campaign streamers for the 283rd Field Artillery Battalion. The War Department acted to correct this error and most personnel records of men and officers include this campaign.

German industrial technology

Fortunately, our distance travel was not hampered by other traffic. There were few civilian vehicles, and those few were propelled by charcoal and wood. Charcoal is usually made from wood. Back as far as 1928 Hermann Goering had planned to use wood as a substitute for petroleum products in their forthcoming war. Douglas Miller, American Commercial Attache in Berlin, wrote numerous books and papers concerning the progress Germans were making in the 1930's toward using wood as a substitute for petroleum. The May 1938 issue of the Journal of Forestry, publication of the Society of American Foresters, included one of Miller's papers. The new products being made from wood included rayon, wood sugars, and Forsmann wood. They had even made a saw blade from Forsmann wood which could cut wood. These products were the forerunner of our 1990's plastics/wood-flour/fiber-glass industry. Other than for heating, the primary use of wood was to propel vehicles. During the war, most all civilian vehicles in Europe and Scandanavia were propelled by wood gas made aboard the vehicle from charcoal. This included large trucks and buses. One cord of air-dried hardwood has the equivalent heating power of 150 gallons of fuel oil.

The Germans also developed technology for converting coal to a liquid fuel, comparable to gasoline. "Using techniques which no other country ever mastered, coal was transformed into synthetic oil so that in 1940 the Germans made about 4.25 million tons of it, of which a great deal was refined into aviation spirit" (Deighton 1993: p.499). "Their 12 plants were producing 32 million barrels per year" (Eagan, 1975;p. 127).

Hedy Lamar, one of our Hollywood queens of that era, provided our government with technical information on some of these developments. She was exceptionally intelligent and could well have been a scientist. Her first husband, Fritz Mandel an Austrian industrialist, had constructed a large wood gas generator in Austria using patents supplied by Goering. Mandel was not a Nazi and fled Austria when the Germans occupied it.

World War II was fought with wood. We have already seen how the United States used wood equivalent to that in 20 million homes. Germany used far more, and without depleting their own forests as badly as they did the forests of conquered nations. Even before the war, in 1936, Germany had been able to buy 2.3 million cubic meters of wood annually from other European countries without spending any of their critically short foreign exchange currency. They had used a system of import quotas and permits along with blatant lying (Armstrong and Oates 1992: p. 27).

114

Standing trees which may contain shrapnel, or large bullets, have little value. If they are felled and taken to a sawmill there is too much probability that the teeth on a saw blade would strike the shrapnel and then some of the saw teeth would become dangerous projectiles. Sometimes logs are screened with mine detectors, but this is costly and not certain. This became a major problem for the European forest products industry after the war.

What if I had been born in Alsace-Lorraine or Germany?

The question has perplexed me ever since the Colmar Campaign. Many Alsatian youth did flee rather than be conscripted into an army which was bent on world domination using the most dastardly means. Two years earlier, some Austrian youth also fled to avoid service in the German army. In addition to the youth many people in numerous countries, as well as Germany, fled to avoid living under tyrranical German rule. However, in Germany very few unaccompanied non-Jewish young men sought refuge elsewhere. Resistance did develop within Germany, including the White Rose Group. However, it was miniscule encompassing but a fraction of one percent of the population. Hence, I always return to the question of why so very few did what I think I would have done?

Born in 1921 means I would have been age 12 when Hitler took over Germany. I would have been appalled to have seen " Crystal Nacht", the brutality of the old Nazis SA Brown Shirts, and the rounding up of Jewish children for shipment to concentration camps. Although the magnitude of the killings within the concentration camps may not have been widely known, the existence of the camps, and the brutality within, was known in Germany and in other countries. With, or without, parental knowledge I would have emigrated before the border-gates were closed. The void, in German youth action, can only be explained by differences in our culture.

The old Germanic world was the Pagan world par excellence. (Bailey 1991;p. 132).

Bailey, in his obsession with Germans, and his lifetime of studying their culture, writes that initially it seemed that

"the Germanic-German preoccupation with carnage and corpses came close to necrolatry and necrophilia, the worship and physical love of corpses and carnification — the production of corpses for the sake of the production of corpses. There is at least some truth in this: it is impossible to separate the hero (Held) from the hero's death (Heldentod) — the hero works with death, as the special troops of the SS had their own strange device: "to give and to receive death." The physical product of death is the corpse (Bailey 1991: p.133).

Chapter 5
COMPLETING THE CAMPAIGN
FOR THE RHINELAND

On To Cologne

In early February 1944 our battalion, having completed the Colmar battle, received orders to support the 104th Division in crossing the Roer River and driving on to Cologne. This meant we were returning to First Army commanded by Lieutenant General Courtney H. Hodges.

The small Roer River should not be confused with the larger Ruhr River. The Roer, although nothing more than an eight-foot deep, 70-foot wide river, had become strategically important because the German forces had selected it for a defense line. They had made the selection because it was generally a south to north flowing river, and not fordable. In Chapter 3, the 29th Division reached the Roer in November. The planned December crossing was delayed by the German Ardennes offensive. Then it had continued to be a relative inactive front until this time. German forces also held a trump card in the system of dams south of Duren. They could release water from these dams, substantially increasing the flow of the Roer River, isolating any bridgeheads which Allied forces might establish, and washing out any bridges the allies had constructed. Allied air forces had repeatedly tried to destroy the dams, but to no avail. British 12,000-pound blockbuster bombs caused no serious damage to the dams. These specific dams were not suitable for skip bombing techniques which were used on the Ruhr River Mohne and Eder dams.

On the 11th of February, First Army was making headway in their drive for the terrain which controlled the dams. This forced the Germans to play their trump card. The dams were so massive they could not be destroyed, so the Germans blew the discharge valves just as we were leaving the Colmar area. Huge amounts of water were released into the partially frozen, snow-covered Roer. The Roer River rose five feet and attained a width of from 400 to 1200 yards. The flow was aggravated by runoff from rain and melting snow. This forestalled any immediate Allied crossing of the Roer.

Supreme Headquarters Allied Expeditionary Forces (SHAEF) had second thoughts about the use of the 283rd Field Artillery Battalion. We went into bivouac at Longechamp, France, after travelling only 80 miles. Our route included Colmar, St. Marie, St. Die, Epinal, and Diginville. After a ten-day delay it was decided to proceed with the original plan because the waters were subsiding. We resumed our northward travel passing through Verdun, Sedan,

Marche, Leige, Aachen, and Eschweiler. By the 22nd of February our howitzers were in position on the west bank of the Roer, across from Duren.

VII Corps and the 104th Infantry Division

The VII Corps, commanded by Major General J. Lawton Collins (known as "Lightning Joe") had landed with the initial invasion forces at Utah Beach, in Normandy. The 104th Infantry Division, whose shoulder patch was a timber wolf on a blue circle, was now in the corps. The division, known as the "Timberwolves", was a reserve division from the Pacific Northwest which had been activated on 7 August 1942. General Terry Allen, who had commanded the 1st Infantry Division in North Africa, was assigned to command the Timberwolves in October 1943. He intensified their training, particularly for night fighting. The division disembarked in Cherbourg on 7 September 1944 and then had an additional six weeks of training in Normandy. Thus, with more than two years in training they entered combat under First Army on the very day I had the encounter with the 35 German officers. German forces had come to the conclusion that Americans only fought in daylight. They were in for a surprise.

The German Forces at Duren, unbeknownst to us at the time

About 15 February, LVIII Panzer Corps under General Krueger, a part of the German 15th Army, was committed near Duren (Dueren). Included were

12th Volks Genadier Div.

353rd Infantry Div.

Artillery Corps 407 committed east of Duren

(Our enemy at Aldenhoven in Chapter 3).

(von Zangen, 1947: p102)

Finally, crossing the Roer River

Our plan called for a night crossing of the Roer preceded by an immense artillery barrage. German troops no longer had the capability of flooding out any crossing. The Germans were aware of our plans and dispatched all available aircraft, including their new jets, to the scene. Hitler's hand was being forced, because he had hoped to retain all his new jet aircraft for one gigantic strike. Allied fighter aircraft had little effect on the jet aircraft. German jet aircraft operated freely. There also were propeller-type German aircraft which were attacked by our fighter planes, and anti-aircraft guns. Aerial dogfights were common in the skies above us.

Initially, we were not the target of the German aircraft. We viewed the German planes bombing and strafing the assembling Infantry as we began engaging targets of opportunity with our artillery fire. Our machine gunners did shoot at the jets with our .50 caliber machine guns, but I don't believe any round ever came close, nor for that matter did any round ever get ahead of a jet. Our 50 caliber machine guns were usually loaded with every fifth round being a tracer bullet which enabled the gunner to see the actual path of the bullets.

At Duren, our battalion fired more than 4,000 artillery rounds in less than 24 hours. The guns were too hot to touch. Stacks of ammunition casings grew to the point that they dwarfed the actual howitzer positions. Brass shell casings housed the propelling charges. The casings were tipped by the projectile. After each projectile was launched the spent brass-shell casing was removed from the breech of the howitzer. They were not re-useable, but were recovered and returned for re-cycling. Manhandling, uncrating, unwrapping, removal of some powder charges, loading, firing, and unloading the spent shell casing of several hundred rounds per each howitzer, all in one day, could only have been accomplished by the sturdiest of young men.

There were 25 types of projectiles for the 105mm howitzer M2A1 model (Sawicki 1977: p.1240). These included fuze quick, time fire, pozit fuze, white phosphorous, smoke, illuminating, shot, and cannister. This variety further complicated fire missions. However, we usually only had three or four types of ammunition at the gun positions.

German jet aircraft returned. This time our artillery positions were the target. The Germans dropped numerous anti-personnel bombs, two of which were direct hits on "B" Battery's guns (Whitt 1945). Then the German jets commenced machine gun strafing of our howitzer positions. We did have a number of injuries, but the men never stopped firing. It was indeed a most memorable scene to watch these young Americans, with an intense sense of purpose, disregard their own safety in order to neutralize German artillery which was causing heavy casualties amongst our Infantry.

Fortunately, our howitzers and other positions were well dug in. When the ground was frozen, as was the case in this instance, Myron Brenner and others used TNT, placed in a crow-bar dug hole, to blast an opening in the frozen ground which could then be expanded by pick and shovel work. I was always impressed with the initative of the gun sections in commencing this arduous work with minimal encouragement from me. Myron Brenner drove a 3/4-ton truck and was our man of all trades.

Before daylight, 23 February 1945, the 104th Division crossed the still swollen Roer. The swift speed of the current caused many problems including some overturned boats and other boats being swept downstream. In at least one instance, our initial bridging cable across the Roer caused an assault boat to capsize when the heavy current washed the boat against the cable in a manner where the upper part of the boat was held tight against the cable, and the lower part of the boat was pushed downstream by the strong current. Several men, weighted down by ammunition, grenades, and other equipment, drowned in the Roer. However, a bridgehead on the east bank of the Roer was soon established.

The next day our artillery battalion crossed the Roer River and wound our way on a winding, recently bull-dozed, path through the rubble that at one time had been Duren. Even the rubble had been pulverized into fine dust. The citizenry of Duren were in for a shock when they viewed the damage their government had wrought.

The drive for Cologne

Eight days of heavy fighting, through towns nearly devoid of civilians, were required to reach Cologne. Our Timberwolves were successful in their night fighting, capturing the Germans by complete surprise on several successive nights, and minimizing their own casualties.

> As we neared Cologne we had the opportunity to view Allied strategic bombing. A large number of Allied bombers flew over us, and shortly we heard the bombs exploding, probably in or near Cologne. The smoke was clearly visible. We observed German Flak guns firing at our planes, but fortunately we didn't observe any of our planes being downed by German fire (Denton 1994).

> The technique of attacking over the open plain by night and mopping up resistance in the villages by day, first exercised to any great extent by the 104th Division, paid big dividends as the advance pressed forward with a minimum of losses (Collins, 1945: p.52).

The largest German city captured by the West-Front Allies

Cologne was the third largest city in Germany. Berlin, the largest, was captured by the Soviets. Hamburg, the second largest city, was declared an open city in the face of the British advance and so was not attacked (Collins 1945:p. 56). Thus, Cologne was the largest city to be captured by the American and

British forces. While the 104th Infantry Division was making the frontal assault on Cologne, our 3rd Armored Division was attacking from the north and our 8th Infantry Division was attacking from the south.

Some of the towns on our Artillery Battalion's route to Cologne included Luchem, Merrzenich, Gut H.S. Forst, Sinsdorf, Konigsdorf, and Weiden. We did encounter one German teenager and his mother in Sinsdorf. We queried him as to why he had not fled deeper into Germany as ordered. His response was "My father was a prisoner of the Americans in the last war, and he was well treated. He advised us to remain in our home." This German lad was healthy and apparently had been fed far better than the thousands of skinny French, Belgium, and Dutch kids we had seen.

German civilians had fled on short notice, leaving behind their livestock, and coops full of chickens. Cellars were well-stocked with coal, potatoes, smoked hams, preserves, and vegetables. This was quite the opposite of what we found in Wales, England, France, Belgium, Holland,and Luxembourg. Most every cellar had a stone-arched ceiling. Apparently they had been preparing for artillery wars for more than a hundred years.

We crossed the Erft River and canal system at Gros Konigsdorf, eight kilometers from Cologne. On March 7, 1945 our gun positions were in Cologne and, once again, we were shelling German forces on the east side of the Rhine River.

> Captain Dopp and I sought out the historic cathedral on that 7th of March. I still have a photo of Captain Dopp standing by a sign "Koln" in front of the cathedral. The roof was damaged more than the other parts (Davis 1993).

> In Cologne, we observed German civilians raiding the large food supply centers. However, this was of no concern to us because we were busily engaged in firing at targets on the east side of the Rhine. They apparently realized that the jig was up and, although up till this time they had had ample food, Germany could no longer plunder other countries to obtain foodstuffs (Denton 1994).

Strangely, the Ford factory north of Cologne had hardly been damaged. Sergeant Finn, and several other men, picked up new trucks off the assembly line. The trucks were shoddily constructed, probably by slaves. They didn't last long. The Timberwolves (our 104th Division) had attained their objective and commenced cleaning up pockets of by-passed Germans in and around Cologne. There was no immediate way of crossing the Rhine. All bridges in that area had been destroyed by the retreating Germans. The Timberwolves would have to wait while assault boats were brought overland from French ports. First Army seized the railroad bridge at Remagen (to our south) on the same day we established gun positions in Cologne. The Remagen American bridgehead on the east side of the Rhine was rapidly expanded despite fierce German opposition. It would be another 17 days before there were other crossings.

The German Saarland

However, several hundred miles south of us the American 7th Army was slated for an offensive into the German Saar and the infamous Ziegfried Line. The Germans called it the "Westwall". This attack would alleviate some German pressure on the Remagen bridgehead and possibly provide one more American crossing of the Rhine River. Thus, our battalion was assigned a new mission, still a part of the Campaign for the Rhineland. Our Cologne gun positions were closed, troops mounted their vehicles, and we were headed south again. Initially, we were to be in direct support of the 44th Infantry Division. Apparently the Supreme Headquarters, Allied Expeditionary Forces' (SHAEF) staff member responsibile for maneuvering the 283rd Field Artillery Battalion was still alert to every opportunity.

Philadelphia recognition

The Philadelphia Evening Bulletin of Tuesday March 6, 1945 cited the battalion for having served with five allied armies. I don't know the source of this news, but some of it might have been questioned by censors. We did not have anyone, to my knowledge, charged with public relations responsibility. Such an assignment would have conflicted with the need for secrecy. The headline read

> "250 Artillerymen from this area in Unit Commended by General."

The article included names and addresses of forty of the Philadelphians. Other Pennsylvania newspapers carried the same article without the names of the Philadelphians. The article referred to the commendation from General F. W. Milburn, commanding the 21st Corps,

The 283rd Field Artillery Battalion was particularly outstanding in these operations (Colmar in Alsace-Lorraine).

The writer also cited some of the operations included in this book. Possibly the source of the news was the January 3rd, 1945 issue of the 29th Division's two-page mimeographed newsletter devoted entirely to the 283rd Field Artillery Battalion.

Their 105mm howitzers are "deadly accurate". They are wielded by an energetic group of men. They are imaginative and enterprising.

The twenty-eight letters, and endorsements, of citation to the battalion are copied in the battalion history. Some singled out individuals, particularly forward observers, for special mention. The rapidity of European Theatre operational movements, along with recurring crises, did not permit most commanders to send the number of citations and letters of commendation to attached units that they would have liked to.

XV Corps and the 44th and 45th Infantry Divisions

Exactly 300 miles, and six days later, we set up our gun positions at Saareinsming, France. Towns/cities we had passed through included Duren, Birksdorf, Haren, Aachen, Liege, LaMarche, Boullion, Sedan, Verdun, Guising, Bemey, Thiacourt, Pont a M..., Baronville, Puttlande, Sarables, and Guising. We were once again with a division which in turn was assigned to the XV Corps. Major General Wade H. Haislip still commanded the corps. The corps was assigned to the U.S. Seventh Army under the command of Lieutenant General Alexander M. Patch. And in turn, the Seventh Army was assigned to the 6th U.S. Army Group, commanded by General Jacob L. Devers.

Initially we were assigned to the 44th Division, a former New York/New Jersey National Guard division. It was commanded by Major General Robert L. Spragins. The very next day we were released from the 44th Division and attached to the 45th Division as direct support. The 45th Division, with a shoulder patch of the yellow Thunderbird on a red diamond, was commanded by Major General Robert T. Frederick. The golden Thunderbird, emblem of good luck and good life, was taken from the mythology of the Native American tribes who lived in the National Guard region of Oklahoma, New Mexico, Arizona, and Colorado. More than a thousand of the original members of the division were Native Americans.

The 45th Infantry Division's three regiments were the 157th, 179th and 180th. The authorized strength of the Division was about 14,500 men. They had many casualties along the way to the German Saar. The Division had landed in Sicily, then into the Italian mainland at Salerno and Anzio, and then they were one of the main forces in the invasion of Southern France. They had fought their way north along the Rhone Valley into the French Vosges where the 157th Infantry was decimated at Reipertswiller at about the same time we were on the way to Colmar (Buechner 1991).

On the 13th of March, 1945 we were in direct support of the 45th Infantry Division. We had no idea what we were in for. Our forthcoming itinerary would have impressed anyone. But, to be in on the capture of Nuremburg, Dachau, Munich, and Berchtesgaden was beyond our wildest dreams. Worms, Aschaffenberg, Bamberg, and Salzburg were comparatively minor cities in our forthcoming Campaign for Central Europe.

Breaching the Ziegfried Line in the German Saar

The Saarland lies north of Alsace-Lorraine, and like Alsace-Lorraine is on the west side of the Rhine. Germany and France have also contested the Saarland for centuries. After World War I, it was given to France with the provision that the people would have a plebiscite in 1935 to decide whether they wished to be in Germany. The Sunday-vote was over 90 percent for union with Germany. This was further indication of the near-unanimous support of Germans for the forthcoming aggression. Shortly after the plebiscite the Germans commenced constructing the heavily fortified Ziegfried Line (Westwall) in the westerly and southern edges of the Saarland.

We crossed the Blies River on a pontoon bridge on March 15, and were back in Germany at Habkirchen. Breaching of Germany's Ziegfried Line was our immediate objective. There were thousands of concrete pillboxes, concrete dragon teeth, barbed wire entanglements, anti-tank ditches, and minefields. The small concrete, mutually supporting, forts had seven-foot thick walls and roofs. Land mines had been in place seven or eight years and were difficult to detect. We lost Lieutenant Gerard Gauthier, the son of Jersey City's Chief of Police. Gerry was reconnoitering a new gun position for "B" Battery when he stepped on an anti-personnel mine. He died in the minefield.

The Battle of the Saar Basin went slowly. German resistance was stubborn, but was eventually overcome by superb infantry action along with our artillery fire. At least one pill box, where the Germans offered considerable resistance, was simply buried by our armored bull dozers. Again, there were heavy infantry casualties, and these were not to be the last. The many fine American infantrymen had been reinforced in early 1945 with a large influx of soldiers who had initially been assigned to study at various universities. When the extent of the American casualties became evident, the decision was made to close down many of the Armed Forces' university programs, and assign the men to the replacment system. This influx of very intelligent young men was of significant benefit to our Infantry forces.

In a few days we were wending our way through rubble which had recently been the City of Saarbrucken. On March 21 we were at the edge of Homburg and realized we had breached the Ziegfried Line. We drove 50 miles to Heidersheim in our dash through the German Saar for the Rhine River. German prisoners were marching west. Dead German soldiers, wrecked German equipment, and smashed pill boxes littered the Saar. We were becoming accustomed to the continual stench of the smoldering remains of buildings, along with the rotting human and animal flesh.

Reconnaissance

Major Fred Dieterle, our Battalion Executive Officer, seemed to pattern his activity along the lines of the Battalion Commander. Neither one of them were command post habituees. Late one afternoon Dieterle requested Pensock as his radio operator on a reconnaissance mission. Pensock enquired about rations and was told they were in the C&R car. Pensock tells his own story:

> As the road went deeper into Germany, our troops were fewer and fewer, till there were none. At dusk we arrived at a small German town and parked about 100 feet from a large weathered gray barn. Here we would break out the rations and bed down for the night. The rations turned out to be a gallon can of cherries, with the pits still in them.

> Dieterle and his driver bedded down in the adjacent house and I, eating the cherries and spitting out the pits, stayed with the C&R running it every hour to keep the battery charged. Darkness settled in and I had only the small red and green lights on the SCR193 transmitter and the BC 302 receiver.

> Two hours later I was startled by a small German boy of about ten years. He spoke softly, quickly and excitedly. All I could understand were the words "German soldaten". Just as quickly as he came, he turned and ran away.

I was alert and scared. Around 11 p.m. the German boy returned and we had the same exchange of words. Again, he ran away. I was relieved to see the first streaks of daylight. Off in the distance, in the still of the early morning, I heard a sound that sent a chill up and down my spine. A tank. But, whose tank? When I saw the big white star of the American 32 ton tank I jumped out, frantically waving my arms for it to stop. The tank stopped, the turret hatch flew open, out came the head of a staff sergeant with the biggest unlit cigar, spouting; "What the hell are you doing here?" I told him we were on reconnaissance and that the officer and driver were sleeping in town.

One man in the tank spoke some German. I spotted the German boy watching us from two blocks away. I walked slowly toward him, so as not to frighten him. I took him by the hand and led him back to the tank. The boy talked with the German-speaking tanker. Then, the tanker talked with the sergeant. The sergeant asked how long had I been here. I told him. "Wait till you see this" he said. The tank turned and moved to within 20 feet of the barn door. The German-speaking tanker spoke loudly in German. The next minute, out came a large number, over a hundred, of fully-armed German soldiers.

The German captain came directly up to me and in perfect English he stated, "Sergeant, we watched you all night". "Why didn't you take me?" I asked. He replied, "We thought that if the American army left one man in a vehicle all night, the town must be full of American soldiers." I asked him if he would like to know how many of us there were. He nodded. I lifted my left hand up holding up three fingers.

The tankers disarmed the Germans and lined them up in front of the tank to march away. Just then Major Dieterle and his driver arrived on the scene blithely accepting credit for the capture. The stogie-chewing sergeant gave a choice reply (Pensock 1993).

Lieutenant Fred Whitt, a forester from Missouri, was one of our two light plane observers. He recognized that German foresters had traditionally planted Scotch and Austrian pine trees on sandy soils and Norway spruce on clay soils. In the wet spring of 1945 our armored forces could traverse the sandy soils.

Hence, Fred was able to assist Army intelligence people by mapping the sandy-soil areas.

The Battle of Central Europe

Thus began the official Battle of Central Europe, sometimes referred to as the Battle of Inner Germany. The Campaign for the Rhineland became history. We now had three stars on our European Theatre ribbon.

Reasons for continued German resistance

Reasons for continued German resistance were difficult to comprehend. Some people felt it was the Morgantheau plan. He was Secretary of the Treasury for President Roosevelt, and had advocated forcing Germany into a position where these things could never happen again. Germany would be divided into small farming states. This plan was one version of the allied call for unconditional surrender. Germany's Minister of Propaganda, Goebbels, used the Morgantheau plan to urge all Germans to fight on. There were other reasons, including German soldiers had few options. Any talk of surrender, any vacating of a military post, any dereliction of duty were dealt with by immediate hanging. Their families in Germany were hostage to their good behavior. Individuals surrendering were considered deserters, their home towns were notified, and their family names were blackened for generations to come. In the closing weeks of the war, roving patrols of German military police apprehended numerous stragglers who were tried and hung on the spot. The eastern front had other motivators for continued resistance, including Russian treatment of prisoners and German civilians. Hitler continually hoped, in vain, for a split among the Allies.

Another incentive for German troops was the pot at the end of their blackened rainbow. Fat Hermann (Hermann Goering) had assured his aerial hit men they would be granted large estates in the conquered lands. Doubtlessly these possibilities permeated into all branches of the German forces.

All along there had been German hopes for deployment of some new secret weapon. The V-I and the V-II rockets had come too late to wreck the allied Normandy Invasion. They were very potent weapons. Even after the invasion of France, Germans still hoped for another more powerful secret weapon. Very few Germans knew the work being done on their atomic bomb, but even they leaked the idea that a war-winning weapon was under development. And, the German people hoped in vain.

The case of the Oslo Papers

Early in the war, plans for the development of German secret weapons were in the hands of British intelligence. The case of the Oslo Papers is still somewhat wrapped in secrecy, primarily to protect the author's descendants from resurgent Nazis.

An unknown person dropped off a package of papers at the British Embassy in Oslo. The British were dubious of the authenticity, but in later years realized they had a complete set of German scientific developments for the war. Included were reports on German efforts to develop an atomic bomb, as well as the V-I and the V-II rockets. The original set of papers has never been released because it would reveal, by handwriting, the identity of the author. The planned 1992 release has been further delayed by resurgent Nazi activity. Speculation has been that the author was Paul Rosbaud, Editor of Springer-Verlag.

One year later, when the Germans occupied Oslo, the Norwegian Jew, Victor M. Goldschmidt, was still there. He was the one person who could have set the German atomic program on the right track. He didn't, and the Germans continued using heavy water and overlooking plutonium. Their Norwegian heavy water supplies were later destroyed by the famous Allied commando raid. The full story is in Thomas Powers 1993 book *Heisenberg's War*.

At the end of the European War, the British assembled many German atomic scientists at Farm Hall in England. The British had secretly installed tape recorders in the German quarters. Thus, when Hiroshima and Nagasaki were bombed the gathered German scientists learned of American success. Their ensuing discussions were secretly audio-taped. Essentially, the German scientists mutually agreed to a uniform position of having deliberately misled Hitler so their atomic program would fail. However, this was not their real position (Powers 1993).

Return to reasons for German continued resistance

German troops were utterly ruthless. They knew that any action of theirs, which was deemed to further their victories, would not be punishable as long as they won the war. French, Belgium, Dutch, Polish, and Czech civilians, and military alike, feared German troops. German troops killed tens of thousands of innocent civilians in thousands of atrocities in retaliation for resistance to their rule. The tens of thousands of German troops involved in these massacres anticipated their own trials, and death sentences, in the event they lost the war. Their only option was to fight on.

Many German troops on the western front had also served on the eastern front, where brutality had been encouraged. Combat with the Russians had necessitated extremely harsh discipline in the German army. There were 13,000 to 15,000 German soldiers executed for their poor combat performance. An additional 427,512 received prison terms. And, thousands of others were reassigned to penal battalions. Surviving a tour of duty with a penal battalion was very tenuous. These battalions were used for exceptionally hazardous assignments. Evading a likely death meant certain execution (Bartov 1991: p. 96).

In consideration of the harsh discipline, German troops were given license to vent their anger on the local populace. German troops were rarely punished and not infrequently praised for rape and plunder (Bartov 1991;p. 93). German officers incited the men to barbarity (Bartov 1991;p. 128-136).

> Once the fighting began, rather than attempt to temper their troops' brutality, many commanders seemed to think that the soldiers were still showing too much compassion for the enemy, and strove to instill in them a greater understanding for, and a firmer will to participate in, the brutalities deemed essential for the outcome of this "war of ideologies." (Bartov 1991; p. 129).

German troops now included teenagers and young men from the conquered nations. Most of the Dutch, Belgium, and even French lads who voluntarily took an unrevokable oath to Adolph Hitler were shipped to the Eastern Front. This made it more unlikely they might be tempted to break their oath. Breaking the oath was, of course, punishable by death. The Netherlands provided 2,559 volunteers for the SS Legion Niederlande (Stein 1966; p. 155).

The motivation for these western volunteers joining the SS has been studied many times since the war.

> A summary of available evidence indicates that the largest group of western volunteers joined the SS for such non-idealistic reasons as desire for adventure, status, glory, or material benefit (in addition to pay and allotments, volunteers were promised civil-service preference and grants of land after the war) (Stein 1966;p. 142).

A substantial number of German soldiers were killed in the last month of the war on the eastern front. These could have been avoided if the German Army senior officers could have swallowed a handful of courage pills and defied the mad man to whom they had sworn loyalty. Again, swearing loyalty to an individual, as opposed to loyalty to a constitution, is fraught with danger. Many of these senior officers were able to save themselves, but were quite willing to condemn their entire class of male teenagers to death, or even worse to being captured by the Soviets.

I had grown up hearing my elders criticize the decision to negotiate an end to World War I, rather than going for unconditional surrender. This time it would be different. The only terms were unconditional surrender. At least, we have had more than fifty years without Germanic aggression.

Our soldiers' newspaper

The soldiers newspaper has been the "Stars and Stripes" ever since the Civil War. During World War II we received daily copies of the European edition in sufficient numbers for each gun section to have a copy. American forces were well-informed, including the Morgantheau plan for German unconditional surrender. Soldiers devoured every word and cartoon. We had ultimate faith in the veracity of every news item. The paper presented both the good and the bad. There were no politically distorted reports. Reporters and editors for the "Stars and Stripes" were definitely on our side.

One case history should be cited. In November 1944, an American Army nurse, named "Sanger" as I remember, published a letter thanking the troops in an emotionally moving way. Millions of American soldiers suddenly became her friend, even her big brother. Less than a month later the "Stars and Stripes" reported her death from German artillery. This one case forged a multi-million member family. We will never forget Lieutenant Sanger.

Chapter 6
THE CAMPAIGN FOR CENTRAL EUROPE

The 45th Division's Rhine River crossing near Worms

The Battalion arrived at Herrensheim, on the west side of the Rhine River on 24 March, and along with thousands of other Americans commenced preparation for the 45th Infantry Division's Rhine River crossing near Worms. The assemblage of 35 artillery battalions to support the 45th Divsision's crossing created the need for a new organization. Our 283rd Field Artillery Battalion was selected to function as an artillery group in addition to the normal role.

About midnight on 23 March 1945 we received general instructions regarding the plan of operation, and were assigned the mission of direct support of the 180th Infantry Regiment. For this mission we were to have the fire of seven attached battalions (listed below), which we were to control and coordinate.

BATTALION	CALIBER
158th Field Artillery	105mm Howitzer
522nd Field Artillery	105mm Howitzer
342nd Field Artillery	105mm Armd. S.P.
189th Field Artillery	155mm Howitzer
182nd Field Artillery	155mm Howitzer
257th Field Artillery	155mm Howitzer
752nd Field Artillery	155mm Howitzer

In addition there were 25 battalions on call (Sapp, 1945).

In the foregoing cited reference, which is on file at Fort Sill, Major Sapp gives the details of the plan including the elaborate communications Traffic Diagrams. Major John D. Sapp, our Battalion Operations Officer, was one more example of persons who had been prepared for such a challenging assignment by adversity. His father had died while he was in high school and he had assumed the lead role in raising his younger siblings. His family was among the poorest, but by dint he was also able to work his way through the University of Florida. He was commissioned a second lieutenant in June 1941 by the Reserve Officer's Training Program (ROTC). His call to active duty came in October 1941.

Lieutenant George Lundgren, the communications officer, worked around the clock to devise a communication plan which would include the 32 additional battalions. He succeeded.

The United States Army's Artillery School, at Fort Sill, Oklahoma, had developed procedures which enabled American forces, in World War II, to mass the fire of many artillery battalions on one target whereby hundreds of artillery rounds would hit at the same time. These were referred to as "time on target" (t.o.t.'s).

We also had the newly developed Pozit fuze available for our shells. It was far better than the previous M-54 time fuze when you wanted the shell to detonate a short distance before striking the ground. The new fuze emitted radio waves toward the ground. When the shell was about 20 feet above the ground the radio waves bouncing back to the fuze intructed it to detonate. Such low aerial bursts were very effective against enemy troops.

The crossing, and the expected breakthrough, would require strong artillery support on the far shore just as soon as possible. Artillery could not wait for a pontoon bridge to be built. An amphibious truck company was assigned to transport our howitzers across the Rhine. The amphibious trucks, called DUKW's, would also serve as prime movers for our howitzers on the far shore until our regular trucks could cross via pontoon bridge. German forces were aware of the planned crossing near Worms. The remnants of the German Luft-Waffe were alerted, but so were American anti-aircraft units and American Army Air Force fighter planes.

The artillery barrage was impressive. Levan Reber was in "B" Battery headquarters. He will alway remember one TOT across the Rhine.

One minute a town and a minute later after the smoke cleared, nothing but rubble (Reber 1993).

Our Infantry crossed in small boats, and then clambered up the steep slippery rock revetments. Skies were teeming with allied aircraft and German jets. The German jets easily outmaneuvered Allied fighter planes. Our artillery positions were strafed and bombed, but our casualties were relatively light. I was the executive officer of "C" Battery, with responsibility for the firing of the howitzers.

Late in the morning of 26 March, "C" battery loaded our howitzers onto DUKW's and started for the Rhine. We had been selected as the battalion advance. The river was far wider than I had thought. I was with the first gun section. When we were halfway across the Rhine, the engine on our DUKW stalled out, and wouldn't restart. We commenced floating down the swift-flowing Rhine (northward) towards the region where Germans still held the east bank. It was near dusk and the poor visibility was aggravated by dark clouds of smoke from the firing and bombing. We did not have radio communication of any type. Fortunately we were spotted just before dark, and a fast moving motor launch rescued us from disaster. By the time we were rescued our DUKW was about two miles downstream from the crossing. We were towed back to the east bank of the crossing site and a mechanic on the motor launch was able to get our DUKW started. We skidded up a landing slot which had been blown in the stone retaining wall. Then we unloaded our howitzer and coupled it to the DUKW. Our infantry had advanced so rapidly we were able to drive the six miles to the old Rhein-Durkheim airfield and our first gun position on the east side of the Rhine. We immediately commenced observed fire, with radio instructions from Lieutenant Clifford Kaufmann, our observer.

> Colonel McDonald and I crossed the Rhine in a DUKW around midnight. The outgoing artillery lit up the crossing site like a 4th of July celebration (Davis 1993).

A new type of warfare

A treadway bridge was completed by combat engineers that evening. This was far faster than had been expected, and amazing in consideration of the width and rapid flow of the Rhine. Armored forces were crossing the bridge as it was being completed. Then the remainder of our battalion made the crossing, and in a few hours we were re-united with our trucks (prime movers). We were now poised for a new type of warfare, a fast moving drive which would deal harshly with any opposition. Batteries were streamlined so only essential howitzers could be where they were needed, with the remainder of the battery following along a day or so later. German telephones were still operating. Numerous burgermeisters telephoned ahead advising other burgermeisters of the penalties to be paid for resistance. Townspeople pleaded with German troops not to resist within the town, or the Americans would simply back off, level the town with artillery, and then drive on through the rubble. There were burgermeisters who didn't get the word, and there were German troops who persisted in following their defend-at-all costs orders. Granted, it probably was a difficult position for some German officers. Disobediance was harshly dealt with. It was hard to realize that, at this stage of the war, Hitler's staunchest support was from front-line soldiers (Bartov 1991;p.144-145). Those towns which offered no resistance had large white flags hanging from every building.

The German populace peered from behind the shuttered and draped windows. Church bells were silent.

We encountered an increasing number of civilians as we drove east from the Rhine. Some were German and obviously had been well fed. There were few, if any, skinny kids. Germans had systematically plundered their conquests so as to provide their own people with the best nutrition. They may well have intended for the children of their conquests to be malnourished. Even the German slaves and prisoners of war had a healthier appearance than the French, Belgian, and Dutch kids.

There was one Australian soldier who had been captured in North Africa and for several years had worked on a German farm. He told me how he was allowed freedom after work hours and could even go to the beer hall. He spoke kindly of the Germans, but then it was obvious to me that he was in trouble. He must have given his parole (agreeing not to escape) to attain such an assignment, and the giving of parole is not loooked kindly on by any army.

Aschaffenberg

The entire Artillery Battalion took up positions adjacent to ours at the old Rhein-Durkheim airport. The airport became our launching pad for the rapid advance through central Europe. Within a matter of hours the Rhine was far behind us. Thirteen miles to Felheim with two gun positions and two German towns obliterated along the way. Twenty miles to Reichenbach, 12 more miles to Grbeiberau, and one more gun position to wipe out enemy resistance. We remained there for a short night. The next day, 29 more miles, numerous towns, several gun positions, a few fire missions, and a new river to cross — the Main at Aschaffenberg. There was fanatical German resistance which needlessly resulted in many casualties on both sides, as well as the complete levelling of the city. Our forward observers accompanied the Infantry in the house-to-house fighting. Combat engineers modified a railroad bridge to provide for vehicular crossings.

Turning to a different information source, other than this narration, the battalion published a unit history at the end of the war, printed by the French and paid for under the historic lend-lease program of President Roosevelt. Page 26 of the Battalion history:

> "The preponderance of enemy time fire didn't prevent our enjoying the spectacle of our air-corps dive bombing the center of fanatical resistance in Aschaffenberg. It took a few days to clear the area; and then we were off again over mountains, through thick forests, and across innumerable canals and streams."

Since crossing the Rhine our Seventh Army had been heading northeast. The mission was to protect the right flank of Patton's advancing Third Army. Of course while 7th Army was protecting Patton's right flank, our 7th Army's right flank was fairly open. The French Army was for the greater part still in the vicinity of the Rhine. This northeasterly direction continued until Schweinfurt at which time it became necessary to redirect Patton's Army or else they would have been operating in territory assigned to the Russians. Patton was directed to veer southeast into Austria. And, as 3rd Army changed course, so did the 7th Army. Thus, as seen on Map #5, at Schweinfurt we veered towards Bamberg and then went on a southerly direction to Nuremburg and Munich. With stripped batteries, travelling well in advance of the Battalion, we rolled on, somedays only 16 miles but usually 35 or 40. Points of resistance were overwhelmed. Our casualties were light.

We regularly saw Allied bombers flying overhead. There was one day when we were watching a large flight that we noticed they were being trailed by a pack of German jet fighter planes. Obviously the Germans did not intend to attack until our aircraft had crossed over the front line. That would allow them to capture any downed airmen. We were able to warn our aircraft through the telephonic/radio network, although the response was that our planes were aware of the situation. Shortly we witnessed the attack along with the downing of at least three of our bombers.

Bamberg

The Battle for Bamberg, and another crossing of the meandering Mainz River, was on April 12. The Battalion history (p. 27) reads

> "Over a pontoon bridge we wormed our way through the night under a heavy enemy barrage, up to the hills overlooking Bamberg. Here the panorama of the battleground was in perfect view. You could see our infantry move in and could watch the time bursts of our own and German artillery. The crackle of small-arms fire echoed across the valley."

Lieutenant Fred Whitt of "B" Battery was a forward observer with the Infantry as they battled the Germans from house to house.

Richard L. Malmberg had been with our fire direction center since the tragic episode at Colmar. A German prisoner informed him of some well-provisioned warehouses in Bamberg. One was supposedly filled with whiskey "Eau de Vie".

We took a jeep and ran across country until we hit a civilian prison camp. Everyone in the camp came to the gate and we asked them where these warehouses were. They pointed across a flat field to four huge buildings which were gray in color. Then the prisoners asked us what they should do. We said they should just get out and go home. In searching through the buildings we had found most everything an Army would need, including whiskey, a room full of false teeth, a room full of knives, and food delicacies from all over Europe. Later I looked back from a fourth floor window and saw a wave of humanity coming from the camp to the warehouse. We got out in time. The people from the prison camp plundered the warehouses.

That night, as we were playing Bridge with a Dutch Sea Captain from the prison camp, we were interupted by some Military Police who were furious because someone had let these people leave the camp. Ostensibly, they had all kinds of diseases and were now roaming the countryside. We didn't tell them that a group of Polish people were having a great party in the next building (Malmberg 1993) (Davis 1993).

There were 22 concentration camps and 165 labor camps in the Third Reich (Bailey 1972; p. 146). The foregoing incident is but one of several similar ones.

A possible case of rape

There were unpleasant incidents, probably the worst was when one of our night-time command post guards reported a possible case of rape of a German girl by one of our best non-commissioned officers. Up to this point he had been the finest of all of our original Pennsylvanians. A great deal of the high morale in "C" battery was due to his pervading leadership. As I remember, the girl had willingly gone into a bomb shelter, but was not willing to proceed with anything beyond kissing. Lieutenant Colonel McDonald, the Battalion Commander, and Martin Williamson, "C" Battery Commander, handled the case so confidentially that even to this day I know very little other than that the girl and her parents were fully satisfied. During the war, our Articles of War allowed death sentence in rape cases. Yet, in view of extenuating circumstances, our Battalion Commander and "C" Battery Commander went pretty far out on a limb without implicating any other officers. Had the matter ever reached higher headquarters, the resulting action could have adversely affected the performance of our Battalion.

The men in our battalion seldom had the opportunity to associate with local women. We were never in reserve, which was where most units engaged in association with local women. Our gun positions in France and Belgium were normally some distance from any settlement. In Holland, our gun positions were close by homes. The Dutch families welcomed the troops into their homes, but they also closely chaperoned their daughters. Travel in Germany, west of the Rhine, was through towns devoid of civilians. For example, in our seven-week gun position at Aldenhoven, by the Roer River, no civilians were ever seen. But most important to the issue, the continual excitement, with each day providing opportunities for some worthwhile activity, subjugated our thoughts about most every other matter. In a normal lifetime individuals would not engage in a worthwhile activity of the magnitude of our daily fare. We seldom talked about our visions for future careers. Nor did we dwell on the past depression era. Our battery did not have any cases of venereal disease. The men were quite normal but they were confronted with a continual 268-day period of history-making excitement. The liberation of France, Belgium, parts of Holland, and Luxembourg were accomplishments of the highest magnitude. Now, we were liberating multi-millions of enslaved children, women, and men. Possibly, we could now understand the jubilation of Yankee soldiers when they liberated the black slaves in the Civil War. Doing something really worthwhile is exciting and does not permit thoughts of personal pleasure. I don't believe there is any parallel to the effect our combat had on human behavior.

Our fast moving drive through central Germany was liberating more than seven million displaced persons. They had been incarcerated within factory compounds which included dormitories. Suddenly, the guards were gone. The slaves, who were tearing down the barbed wire enclosures, were not being shot. Freedom had arrived. Some of them wanted immediate retaliation, whereas most wanted food and decent shelter. Looting of German residences by these displaced persons could not be stopped, and furthermore that was not our mission. The number of civilians we encountered increased significantly as we drove east. Sometimes it was difficult to differentiate between Germans and displaced persons. Official regulations forbade American troops fraternizing with German civilians. It had little effect on us because our dominant objective, of capturing all of Germany as swiftly as possible, overwhelmed all of our activity.

Nuremberg and the close-range direct-fire artillery duel

Four days, numerous towns, and 65 miles later we drove into Fischbach, a small town seven kilometers southeast of Nuremberg (Nurnberg is the German spelling). We had encircled "Hitler's City" from the north and east. The German troops who were holed up in the "Old City" were under orders to defend at all costs. The following case history is scarcely believable today, so initially I will quote from the official Battalion history which was published in 1945.

On the 17th April we were on the move again; and finally, after fighting nothing but dust arrived outside Nuremberg. Our arrival at Nuremberg (Fischbach) was given a warm reception by the Germans. The advance elements streaming down the autobahn passed the infantry on tanks, and moved up with the forward foot-soldiers of our combat team, who were halted at that time by a hidden German machine-gun. Men in the lead battery jumped off the truck to get under logs for protection, while the infantry accomplished their mission. Finally the machine-gun was silenced by fire from another artillery battalion and we moved out to take up positions in Fischbach.

We had been in position scarcely ten minutes when we received our first shelling—time fire—from German 88mm AA guns. The bursts were too high and deflection was off, but we knew it would not be long before that was corrected. Rear elements of the battalion coming up the autobahn an hour later received strafing and bombing from the vanishing Luftwaffe, but suffered no serious casualties.

When all the Battalion was in, "A" battery reported they could see, from their position, the German 88's that were firing on us. They fired a few direct rounds but then darkness settled in and things quieted. Early the next morning tanks moved through our position followed by infantry. Then again those 88's opened up. We received clearance to fire and it became an artillery duel. We suffered many casualties, but when it was over found we had knocked out eight of the 88's and one multiple 20 mm gun, as well as German dead, wounded, and captured. When our infantry had passed through we found that altogether there were eighteen of the 88's well hidden and camouflaged.

The German 88's were at the west edge of Fischbach and just south of Fish Creek and the road parallel thereto. The location is marked "SL" on Map #3. Their aircraft warning units, which played no part in the battle, are indicated on the map by "SW". Their mutually-shared barracks were along the westerly side of the autobahn and marked with "D" (Geiger, Burgermeister of Fischbach, in letter of October 5, 1993). Map #3 is an enlarged 1943 U.S. Army map, with gridlines at 1,000 meters.

Our firing batteries moved into position shown on the map by "A", "B", and "C". We had encircled Nuremburg and so were approaching from the northeast along the Autobahn. We veered off the Autobahn towards our gun positions without actually going into Fischbach. It was near dusk and the German 88's were not initially discernible across the well-wooded creek line. The creek was not fordable.

The near-dusk, 17th of April, aerial bursts from the 88's were roof-top height as attested to by the action-photograph taken by John Rearigh and pictured on page 27 of the 1945 Battalion history. The resulting shrapnel inflicted some wounds on our men and pierced many of our vehicles. Many of the German 88's could be seen from our "A" Battery gun position. These were the earlier model high velocity guns which had a high profile, making them quite vulnerable to our howitzer fire. Special versions of the gun, for anti-tank useage, had much lower profiles. During the night our gunners dug in and prepared for a real battle. Some of the support personnel took refuge in the basements of the four new, steeply-roofed, three-storied homes. Our infantry, along with supporting tanks, continued to move into position in the vicinity of Fischbach during the night.

At the same time, the German troops were demolishing their parapets so that they could fire "fuze quick" bursts at us directly. The range was 700 to 900 meters. There was a line of Lombardy poplar trees, just breaking into bud, along the road parallel to the creek. There also were trees and brush in the creek line. At daylight we resumed fire. The Lombardy poplar trees intercepted a few rounds creating havoc with the very low airbursts. The trees were soon shredded creating an open field of fire for the dueling parties. Now, the 88's could be seen from all battery gun positions. The ground surrounding our howitzer positions was virtually plowed up as German shells struck even into shell holes. Lightning may not strike twice in the same place, but artillery shells do!

Many German shells hit "A" Battery which was nearest to them. Maurice London, of Philadelphia, was hit in his mid-section by one 88mm shell. He was survived by wife and children. Six other "A" Battery men were seriously wounded. We fired "fuze quick" which meant the rounds detonated upon impact. We could not use "time fire" aerial bursts with either the new "Pozit fuse" nor the M54 fuze because of the close range. Several German 88's took direct hits from our 105mm shells with devastating results for the German gunners. Shortly, the German fire slowed and a few white flags could be seen. Our Infantry immediately moved into the area and rounded up the prisoners. Some of them were young boys. There were no girls at this German Flak Unit. German anti-aircraft units were a part of the German Luftwaffe (Air Force) and not a part of the Wehrmacht (Army) as in U.S. forces.

Treibspiegelgeschosse were used by the Germans in the Ardennes Offensive. It is not known whether they were used in the Nuremburg artillery duel. They are devices resembling a disk, attached to a projectile to increase air resistance. A projectile fitted with such a device and fired from an 88mm antiaircraft gun would have a curved trajectory resembling that of howitzer ammunition (Information obtained from Genobst Guderian) (Schramm 1946: pp.228-229).

This amazing success, in twelve 105 mm howitzers knocking out eight of the highly vaunted 88's, and capturing ten additional, is probably an all-time record. There were direct-fire artillery duels in the Civil War, but there were not very many on the western front of World War II. This is the only one the author has located in his research.

Nuremberg's Old Walled City and ever onward

We immediately turned to more pressing matters. Our infantry was advancing rapidly past the German 88's toward Nuremberg. There were numerous pockets of German resistance and our artillery support was direly needed. German forces in the "Old Walled City" part of Nuremberg held out for four more days. Rather than endangering many Americans this late in the war with an assault, the 45th Division relied on artillery to subdue the German holdouts.

Still, there was more to come. It just is a matter of the way the cards fall. The battle for Nuremberg was intense. We fired thousands of artillery rounds into the old walled center of Nuremberg. I did make a side trip to the empty Nuremberg Stadium. I, and the men with me, each acquired an engraved stopwatch as a token of having captured Hitler's city. The surviving German troops surrendered and on 22 April we resumed our drive. The word was that Hitler, along with his best troops, were planning to make a last stand in the well-fortified underground bunkers at Berchtesgaden.

> Near Nuremberg, Markowitz and I were searching a four-story school building. On one of the floors we saw a little man come out of a room and run down the hallway. We gave chase, up and down stairs, in and out of rooms, but we couldn't catch him. On the top floor two shots were fired. When we got to the room where we thought the shots had come from we cautiously entered. A woman was half on and half off a couch. She had been shot in the chest. The little man was in a chair behind a huge desk, the upper part of his body across the top of the desk. He had shot himself in the mouth (Parker 1994).

(Ed. comment: Many Germans were so caught up in the Third Reich that they had no wish to live any longer when they realized the jig was up.)

The 42 miles to Georgesmund was completed on 22 April. Then Kattenhochstadt, Rahlingen, Amersfeld, Etting, and the crossing of the Danube were finished by 27 April. Our main objective was Munich!

Dachau, April 29, 1945

The 157th Infantry Regiment of the 45th Division liberated the infamous German concentration camp at Dachau on the 29th of April in accord with assigned orders. The Battalion Commander, Lt. Colonel Felix L. Sparks (Later Brigadier General and Justice of the Colorado Supreme Court) tells his story as written by the wife of the first American medical doctor into Dachau (Buechner 1991: p. 141).

> As the main gate of the camp was closed and locked, we scaled the brick wall surrounding the camp. As I climbed over the wall following the advancing soldiers, I heard rifle fire to my right front. The lead element of the company ("I" Company) had reached the confinement area and was disposing of the SS troops manning the guard towers, along with a number of vicious dogs (Buechner 1991).

> At dawn on 29 April I was with "I" Company of the 3rd Battalion of the 157th Infantry. The Company came upon 50 to 60

box cars on a railroad siding adjacent to Dachau. The cars were ladened with about 2,500 dead bodies. Some had died enroute whereas others had been shot through the head. Our troops became boiling mad after seeing such cruelty. They advanced into Dachau without any regard for cover or concealment and killed all the guards in the watchtowers as well as other SS troops. Many other atrocities came to light before their unbelieving eyes. They found 32,000 inmates, living and dead, in the compound. There were 2,000 to 3,000 naked bodies stacked head high against the walls enclosing the gas chamber. Their clothes were piled in the center of the yard. Naked bodies were also piled against the walls around the ovens awaiting cremation. Gold and silver fillings had been knocked from their mouths. The stench of death was overwhelming and unforgetable (Harvey 1994).

Captain Arthur E. Harvey, Jr., was intelligence officer (S-2) of the 283rd Field Artillery Battalion. He frequently functioned as a high ranking forward observer. A great deal of the favorable reception our infantry forces gave our battalion was due to Arthur's calm demeanor and his superior intellect. This descendant of the great Choctaw Indian Chief, Pushmataha had been raised in the swamps of Burnt Corn Creek in Alabama.

Later that day, three jeeps from the 42nd Infantry Division, carrying a Brigadier General Linden, Maggie Higgins (the reporter), and photographers, arrived to stage a liberation scene (Buechner, 1991). Over the years there continues to be disagreement concerning who liberated Dachau. It is important to set the record straight and honest. However, the main point is that the U.S. Army liberated Dachau. Our Artillery Battalion was in nearby Feldmoching.

"At Feldmoching many of us went to the notorious concentration camp at Dachau. Viewing it we realized we were fighting a nation of fiends; that the German people could never recompense the world for the horrors perpetrated there." (McDonald 1945: p. 28).

"Visited Dachau, a German concentration camp. Hundreds of dead bodies in railroad cars and piled in rooms ready for cremation. The most horrible crime committed against humanity that it's not even possible for a human being to conceive of such a thing to have happened. (32,000 prisoners still alive)" (Whitt, 1945).

Our sense of purpose expanded as we drove into the European fortress. It still focused on the protection of our long-run national security. However, in Wales and England we had seen the results of German bombing. In France we learned of the 4-year enslavement of more than a million Frenchmen and the confiscation of food and other material. In Holland we learned of the forced labor required of all Dutch 18-year olds. In Germany we observed the many displaced persons who lived and worked in appalling conditions. And, now we had seen Dachau. Anyone who tells you there was no "Holocaust" is lying. We saw it. Retribution was now a firm part of our purpose.

The Battle of Munich

The Battle of Munich, 30 April 1945, was quickly settled. Most of the German resistance was from the small towns around Munich. "B" Battery fired on German troops trying to escape by rail cars. These were the last rounds we were to fire in the war. We then wended our way through rubble-strewn paths which may, or may not, have formerly been streets. We were assigned a new mission, by radio, as we drove into Munich. It called for our expeditious movement through the city, and onto Berchtesgaden. We were to join a task force which hoped to reach the Berchtesgaden area before any German resistance could assemble there. Allied forces were not fully aware of the extent of tunnels and underground facilities which may have been prepared to enable German forces to hold out for weeks or even months. The worst-case scenario was assumed. Our convoy revved up engines, and with a few horns blaring from our gun trucks raced through the main parts of the old city. Drunken German soldiers, many still armed with rifles and grenades, silently waved us on. The acceptance of their surrender was not our affair. Civilians, recognizing the urgency of our movement, moved well to the side. The 45th Division remained at Munich to accept surrender from the thousands of German troops.

Evidence of a planned German last ditch stand

Our intelligence had some evidence that the Germans were planning a last-ditch stand in the vicinity of Berchtesgaden. For most of the war the U.S. Army, and the U.S. Navy, code breakers had been deciphering the messages which the Japanese Ambassador to Berlin had been sending to Tokyo. The Embassador, General Oshima, "appeared completely trustworthy to the Nazis, and only he could get close enough to Hitler to learn firsthand what the Fuhrer was thinking. Throughout the war Oshima revealed the Fuhrer's thoughts and plans...". (Boyd 1993; p. 178). Thus, Allied Forces learned that Berlin's foreign embassies were moved to Bad Gastein (50 miles south of Salzburg) in early April 1945. Even more relevant, Oshima reported to Tokyo that the German government planned to move south (Boyd 1993; p. 171).

Hitler takes the easy road out

Hitler spurned the idea of escaping from Berlin solely because he didn't want to chance dying in the Berlin streets like a dog. His alternatives narrowed rapidly as Russian forces closed in on his bunker. Thus, at the same time that we were racing through Munich, Hitler was committing suicide. His disenchanted troops had a most difficult time rationalizing why their leader, who had vowed to fight on to the death, had not done so. He had taken the easy road out.

Continuing with the XV Corps, but the 42nd Infantry Division

The SHAEF staff member responsible for our assignments was not going to miss a beat. We were now assigned to the 42nd Infantry Division, commanded by Major General Harry J. Collins. They had landed in France in January. The Rainbow Division had a rainbow for their shoulder patch. This signified that they had been a national guard division representing all states. We joined the 42nd Infantry Division on the autobahn. All four autobahn lanes leading southeast from Munich were packed with American tanks, mechanized infantry, and artillery all flowing easterly. The median strip was filled with German prisoners, not under any guard, heading west. This scene is frequently recalled by many of us when we drive along our interstate highways. This vivid portrayal of the end of the war was a truly dramatic scene. For us, it meant we had clearly beaten the most evil armed force the world has ever known. No, our interstate highways were not patterned after German Autobahns, but German Autobahns were patterned after New Jersey's Black Horse Pike and White Horse Pike.

Detours were necessary where bridges had been blown. Abandoned German jet aircraft were all along the autobahn, concealed in forested areas. They had been using the autobahn as a landing strip. One regiment of German troops was in parade formation in a field adjacent to the autobahn. Their Colonel was waiting for some American to accept their surrender. No one had the time. Our objective was to capture any redoubt area which the Germans might use. On the 3rd of May the 42nd Division was directed to the north of Salzburg, Austria. The U.S. 3rd Infantry Division would target Salzburg. They would need the 283rd Field Artillery. Thus, with only two days remaining in the war, our battalion was once again assigned to support the main and final offensive.

XXI Corps, and the 3rd Infantry Division

We had been with General Milburn's XXI Corps at Colmar. This was a return engagement. The 3rd Infantry Division, commanded by Major General John O'Daniel, had fought the Germans in North Africa, Sicily, and Italy. They were in the August southern France assault with the 7th Army, and also in the Battle of Colmar.

Salzburg, and Berchtesgaden — our Battalion Command Post

Perlach, Essernef, Grasau, Mauthen, and finally Salzburg. Supposedly, the German SS were defending Salzburg. However, Salzburg fell to the 3rd Division on the 4th of May. Our final gun positions were in Salzburg. The word was the war was about over, but not before the 283rd Field Artillery Battalion evinced one great initiative. Quoting from the Battalion history (p.29)

> Our command post made a dash to Berchtesgaden and set up our final command post in Hitler's mountain retreat (McDonald 1945).

Sergeant Walter L. Scott, our sergeant major, and his driver Bill Davis were in the advance party (Davis 1993). Major Dieterle was probably the first American to enter Hitler's retreat.

> When we arrived at Hitler's retreat I looked around for a souvenir. My eyes swung around the main reception room and there on the fireplace mantel was a two-foot high hand-carved mare and foal. I picked it up and carved on the underside was:
> Dem Fuher
> Adolf Hitler
> Zugeeignet Marz 1933
> Gottlieb Kottmann — SA
>
> (Dieterle 1994)

The sands of time had run out for the early Nazi art depicting the parental relationship between Hitler and the German people. Gottlieb Kottman's Nazi art, representing Hitler and his charges, had fallen into American hands. This was the mare, who uncannily had forehead hair like Hitler. Hitler was so pleased with this gift, which denoted him caring for the German people (represented by the mare's foal) that he had a postage stamp made depicting the art. The late Adolph Schickelgruber never dreamed it would fall into American hands. The Third Reich only lasted a dozen years, but even that was far too long.

Awards

V-E Day was May 7th. The Campaign for Central Europe was over and most members of our Battalion eventually received a fourth star for their European Theatre Ribbon.

Awards for heroism were highly regarded during the entire war. At the time of the breakout from Normandy I recollect some one pointing out an infantryman from the 30th Division and stating he had earned a bronze star for heroic action. He was the only soldier in his infantry batallion who had an award at that time. Bronze stars were meaningful recognition of men's sacrifices for the good of their outfit.

Our Artillery Battalion, like most other units, only made awards when they were truly warranted. After the war, when we had moved back towards Marseilles awaiting shipment home, the record for our battalion was

Bronze Stars	— 37 including oak leaf clusters representing second awards.
Air Medals	— 11 including oak leaf clusters.
Certificate of Merit	— 9

And for the wounded and KIA — 93 (approx.) Purple Hearts.

When our seriously wounded were evacuated through other aid stations than our own, and when these wounded did not return to the battalion because of the serious nature of their wounds, we had no record of their being awarded the Purple Heart. I estimate there were 23 such cases. The battalion had records for only 70 Purple Hearts.

There were no higher awards, such as Silver Stars, Distinguished Service Crosses, nor Congressional Medals of Honor. However, when the Department of Defense recognized award stringency in most of the armies involved in Europe they did make further awards based on letters of commendation which had been issued during the war, and also on proven instances of heroism. For example, on 14 December, 1944 the commanding officer of the 125th Mechanized Cavalry Squadron, Lieutenant Colonel Kleitz, wrote a letter citing eleven of our Battalion's officers and men for heroic achievement. Eventually, they all received Bronze Stars.

There was a "V" device attached to Bronze Stars standing for "Valor". Bronze stars, without the "V" device were usually for meritorious service.

British and Canadian forces may also have been overly restrictive on battle awards to front-line units (Whitaker, Whitaker 1989: p. 203). Whitaker states that for every British/Canadian combat award that was made, one was also made to the headquarters based men. However, the Soviets and Germans were apparently most liberal in awards. The Germans awarded about 2.8 million Iron Crosses of various types (Deighton 1993: p. 603).

> "If you saw a Russian soldier without a decoration of some kind on his left breast" said the American war correspondent, Walter Kerr, you could be sure he had not seen much action" (Deighton 1993:p.483).

Promotions

Promotion possibilities were very limited. Our Table of Organization and Equipment (TO&E) specified the maximum rank for each position. A few promotions were available after our unfortunate fire-direction center deaths at Colmar. Most of the men, who never attained rank beyond Private First Class, were exceptionally good soldiers. They realized our promotion limitations and didn't complain. Likewise, officer promotions were also constrained by the TO&E.

There were 29 officers in the Battalion. Our authorized officer strength, beyond which we could not promote, was

1 lieutenant colonel

2 majors

9 captains

12 first lieutenants

5 second lieutenants.

Those officers who elected to remain in the Army Reserves, who had more than one year in grade, and who were recommended were promoted one grade upon separation from active service.

Ever Onward

The Battalion returned to Munich for a few days and then was assigned a new, and most important mission: Operate the Prisoner of War camp at Fuerstenfeldbruk, 21 miles distance from Munich. We would have 65,000 prisoners in what many years later was the infamous massacre site of the Jewish Olympic Team.

Chapter 7
AFTERMATH OF THE WAR

Germany immediately after surrender

At the end of the war there was no German government. The conquering powers became the government in their mutually agreed upon zones. The European population upheaval defied belief. In the American zones of occupied Germany and Austria there were more than twenty million homeless persons, along with five million disarmed German troops. This included 100,000 Jews who had survived the holocaust; 7,000,000 displaced persons who had been enslaved in Germany; 1,600,000 displaced persons who had been enslaved in Austria; and 12,000,000 ethnic Germans who had been expelled (or fled) from Eastern Europe. It was widely expected that this massive chaos would be compounded by former German troops engaging in guerilla warfare against the occupation forces (Werewolves was the code name). From all parts of the world there was clamoring not to allow the culprits to escape punishment. And, there were the Yalta agreements requiring immediate repatriation of Soviet citizens.

The Wehrmacht (German Army) infra-structure no longer existed. The German railroad system was in shambles. There were few, if any, truckers, only sufficient gasoline for American needs, and few operating highways. Hospitals were overflowing with badly wounded patients. There were millions of maimed persons, no longer able to engage in economic activity. Many of the German young men, born in the period 1918-1924, were dead. Food, medicine, vehicles, appliances, fertilizers, and most everything else were in critically short supply. And, after Hiroshima and Nagasaki there were tremendous demands to get American soldiers home.

Streets of many towns and cities were piled high with rubble. There possibly was a narrow, winding one-way passage through some of the former roads and boulevards. Bridges and viaducts had been destroyed. The stench from burned out buildings, and rotting human and animal bodies was everywhere. Cities were paralyzed. Survivors had taken up residence in, and under, the ruined buildings. The victorious Allied Forces had requisitioned many useable buildings. Allied military governments were well prepared for such chaos and in a matter of days the German people were hard at work clearing the debris. In Munich, the rubble was piled high on one plot of ground creating a new mountain which eventually became a park.

For Germany, 1945 was "Year Zero," the dawn of an even more extraordinary time than 1919 had been. In the first place,

Germany — the whole of it — was occupied by the victorious Allies coming from East and West. In the second place, the Allies came, however ambiguously the fact and the term were interpreted as "liberators" of the Germans from the long night of Nazism. In the third place, the population of Germany as a whole had suffered far more during World War II than during World War I. The latter German population was in a state of shock: the Germans were benumbed, totally disoriented, and psychologically dishelmed. In the fourth place, thanks to the almost untellable atrocities of the Nazis, the Germans found themselves in a state of pariadohm unprecedented in the annals of history (Bailey 1991: pp. 439-440).

The Germans, even though living underground, dramatically assumed prosperity. Women went to work in high heels, coats, and fur pieces. White-collar men wore suits and hats. Even those men clearing rubble (which included numerous diplaced persons) were neatly attired. It was tantamount to a Wagnerian opera, with everyone putting on their best face, perhaps to deny their loss of another World War. Whereas most had been well fed during the war, now they were existing on less than 1,500 calories a day. German families fed their children as well as possible. But then, there were many orphans. Pensions for former military officers were discontinued. Generally, the pensions resumed in 1951.

Many displaced persons, especially from Poland and Czechoslovakia, decided to remain in Germany rather than return home where the Soviets now ruled. Most of them forgot about retribution and decided to become good German citizens. They were living in tent camps erected by the Military Government for Displaced Persons. American soldiers had been forbidden to fraternize with Germans ever since the Battle for Aachen in October 1944. The non-fraternization began to break down when GI's naturally became friendly with the children. It seemed the children really admired the Americans. The non-fraternization policy was dropped in July 1945.

There also were 425,000 German and Italian prisoners of war in America who were slated to be shipped back to Europe as soon as possible. Prior to October 1944, German and Italian Prisoners of War captured by Americans were shipped to America where they were accorded every privilege of the Geneva Convention. Some additional shipments continued through April 1945.

Miraculously, the anticipated disease epidemics never materialized, probably due to the DDT dusting of millions of refugees as they crossed the larger bridges. Fortunately, there was no guerilla warfare, and the vast majority of Germans cooperated with American forces. The German monetary system was able to continue for a few years before it had to be revised. The value of the German mark was pegged at ten cents in 1945.

Our new assignment, points, and transfers

On May 14 our battalion was assigned responsibility for the Disarmed Enemy Forces camp at Fuerstenfeldbruk. Fuerstenfeldbruk was a small German town 20 miles west of Munich with a major Luftwaffe air base. Our airforce had repeatedly bombed the airbase. At the end of the war there were some intact facilities. The runways were deeply cratered.

The reason for the assignment was a combination of our having been through four campaigns whereas units with few campaigns were slated for immediate assignment to the Asiatic Theatre where the war continued. There were exceptions, such as the veteran 45th Division which was slated for the Far East via the United States.

Ship transport priorities were for units being deployed to the far east. We, being slated for return to the U. S., had no shipment priority. This was further complicated by the point system whence our men, with low points, were transferred to units going to the far east. And, in turn, we received transferees who had accumulated high points in other artillery battalions. Some of our original troops had sufficient points to return home in July 1945. The required points for discharge were initally 85, but this was dropped to 80 in September, 60 in October, and then to 50 in December of 1945.

		Typical G.I.
Points were derived from a formula:		in 283rd FA Bn (July)
Months in service	= 1 point each	25
Months overseas	= 1 point each	13
Campaign stars	= 5 points each	20
Combat decoration	= 5 points each	
Children (max of 3)	= 12 points each	_____
		58 points

Disarmed Enemy Forces at Fuerstenfeldbruk

The large numbers of German prisoners, confounded with the shortage of food, further complicated by the largest population upheaval the world has ever known, made it impossible to feed and house captured Germans in accord with the Geneva Convention. Accordingly, they were not prisoners of war, but disarmed enemy forces (DEF's).

The administration of Fuerstenfeldbruk Disarmed Enemy Forces (DEF's) camp was eased because internal management of the camp was performed by a German colonel and his staff. The colonel was fluent in English having been the German military attache in Chicago during the 1930's.

We had 65,000 prisoners in a large field which had been fenced in and then partitioned into 14 sub-compounds. There were no buildings within the actual compound, but it was May and the vast majority of prisoners were to be released within a few weeks. Many DEF's had small tents. Additional prisoners continued to arrive. Most came in willingly because they wanted their release papers, and this was the only method.

The vast majority were German which included some Alsatian renegades. However, most of the Alsatians in the German Army had been on the eastern front, and were either dead or in Siberian prison camps. Few of the Alsatians who were Soviet prisoners ever returned. We had a large contingent of Russians who had fought for the Germans under the infamous Russian General Vlasov. I would estimate their numbers at 3,000. I don't remember any Italian, Romanian or other nationalities. However, differentiation was difficult. Many foreigners, particularly ethnic Germans, had been accepted into the SS and the Wehrmacht. They were granted all the rights of German soldiers, and for all purposes were Germans.

> In 1943 the Waffen SS was compelled, through heavy losses, to relax its special selectiveness and large numbers of foreigners were recruited or conscripted. By the end of 1944 more than half the Waffen SS were not native Germans. They numbered in their ranks men from both East and West Europe and even from Asia (Beck, p. 52).

> Many ethnic Germans from other countries and foreigners from all over Europe served in these units — Especially the Waffen SS (Bischof and Ambrose 1993:p. 129).

The partitioned compartments included one for general officers, gauleiters and other higher echelon Nazis. Gauleiters were Nazi party chieftains of states, such as Bavaria. There was one compartment for other officers; one for SS; one for the women's auxilliary; and one for the Russians who had fought for the Germans. We had none of the Volksturm elderly nor children. They apparently had been sent home. We had no seriously wounded.

> When we first assumed responsibility for the disarmed enemy forces at Fuerstenfeldbruk, the German colonel in charge of camp administration informed me of a serious problem. We met at the entrance to the compound where we had the German generals. Fifteen German generals supported the colonel in stating that if we did not remove a certain obnoxious SS general, they would not be responsible for any harm that came to him. I arranged to have the SS general transferred to another detention camp that day (Dieterle 1994).

The German colonel had his own staff, and police force, to patrol the wide lanes between partitions. He was charged with food distribution along with discipline. Prisoners were fed two meals a day the food being delivered by German civilian authorities and the German Red Cross. Much of the food was American 10 in 1 rations (Witt 1945). Men from our battalion manned the perimeter guard towers.

One of the eight interior cages had been set aside for the uniformed women who were a part of the German Army or the SS. There were only 55 of them. One sunny day they created a crisis when they sunbathed entirely in the nude. It caused a riot as the soldiers in neighboring cages flocked to the fence lines. We were compelled to take immediate remedial measures. They were moved out of the cage and into one of the few surviving buildings on the airbase.

Medical and dental facilities were available in Luftwaffe buildings close by the compound. These were staffed by German Army medics. They were short on medications, and especially pain killers. Dental extractions were without novacaine. A dentist, Dr. Leonard Friedman, had recently been asigned to our Battalion. He was charged with inspecting the German facilities as well as our own medical/dental problems.

Many of our officers, including myself, were in charge of the camp on occasions. Our post was in the main guard tower. One of our problems was keeping civilians some distance from the wire fence. I don't remember any deaths, nor seriously wounded. American officers could sign out prisoners for work details. Most all prisoners were anxious for such assignments. The chance of getting out of the cage for a day, along with an additional meal, was prized.

I frequently viewed the sea of humanity from the guard tower, pondering on their, and my own, future. Few of these Germans had any profession or trade. Ever since their initiation into the Hitler Youth their education had been funnelled into becoming more adept at killing people. It was obvious that they also were concerned about how they would now fit in and be able to make a significant contribution to the economic system. Few, if any, realized that because of so many deaths amongst their male counterparts, those who were physically and mentally fit would be very much in economic demand. I had also grown up in the Army. Overall, these thoughts led to my determination to attain a college education as soon as possible. It would probably be in forestry. About 25 May the American CID (Counter intelligence) people came in and started screening each prisoner. All were released except officers, gauleiters, SS, and Russians. SS could be easily identified because they had their blood type tattooed under their arms.

Colonel McDonald had become attached to a young man who claimed to be a Russian. He became the Colonel's orderly and there even was talk about adoption. The CID interrogation proved that he was a member of the SS by the blood-type, tattooed under his arm. He could have been both SS and Russian as just cited from Philip Beck's *Oradour, Village of the Dead.* He was returned to the cage.

Russian officers interrogated our Russian prisoners the last week of May. They were removed from the cage in small groups, sometimes forcibly. Once they left the compound they never returned. I understood they were being shipped to Russia where many faced execution. My antipathy towards German forces, including all of their allies, had been reinforced by the tour of Dachau. The forcible repatriation of these Russians aroused no sympathy from me.

During this period there were many Germans advocating that the U.S. join with Germany in repelling Communism. At one point a rumor, that the U.S. was going to repel Communism and would soon be asking for volunteer Germans, swept through our DEF camp. It may have been intended to delay the repatriaton of the Russian prisoners. This rumor seemed ludicrous to Americans. Nevertheless, it seemed to be accepted by Germans. Bartov (1991) writes about this wild idea on pages 141-142.

The subject of the treatment of Disarmed Enemy Forces has recently re-surfaced because of a Canadian writer (Bacque, James. 1989. *Other Losses: An investigation into the mass deaths of German Prisoners of War at the hands of the French and Americans After World War II*).

Gunter Bischof and Stephen E. Ambrose (ed.) have refuted each of Bacque's charges in *Eisenhower and the German POWs — Facts against Falsehood* 1992. Lousiana State University Press.

Forward observer teams
rewarded with travel

Captain Martin Williamson, who commanded "C" Battery, suggested I load up my jeep and trailer with gasoline and rations and along with Jacques Linn, my driver, and Leonard Jannsen, my radio operator, do some travel. It was a great experience to be free to travel in a very self-reliant mode. We had no passports, but we did have our rifles. We first toured the shores of Lake Constance. Our evening accomodations were abandoned German Army hostelries and sometimes, mansions. Next we attempted to drop in on Switzerland, but the border guards stated, where as we were welcome, we would have to leave our jeep at the border post and walk in. The cafe would not accept our only money (German Occupation Marks). They wanted American dollars. Their eventual compromise was they would feed us without any charge. The Brenner Pass from Austria to Italy was next on our agenda. At the north end of the pass we found an abandoned German Army rest and recreation hostelry. The buildings were well furnished and undamaged. The kitchen was as modern as any to be found at that time. Some Austrians told us how the elite of the German Army had vacationed there with selected German women. When the German women became pregnant they were rotated back to Germany to bear a new generation of elite people. We returned to Fuerstenfeldbruk via the heavily bombed Innsbruck.

We were relieved of our prisoner assignment on May 30th, by an American military police unit. Many of the German Wehrmacht had been released by this time.

In 1955 I was an Engineer officer in the Army of Occupation. On a Danube River crossing exercise I was assigned a company of German Labor Service who operated DUKW's (amphibious trucks). The commander was a Lieutenant Clum who rememberd me very well, although I had not known him amongst 65,000 prisoners. We talked about the Fuerstenfeldbruk cage and it developed that he had been one of the SS. He claimed he had been very young and hadn't realized he was selling his soul to the devil. He agreed the prisoners had been as well cared for as possible under the circumstances.

General Patton visits our prisoners

General Patton was in command of the American Occupation forces in Bavaria. The following quote is from *The Last Days of General Patton* by Ladislas Farago, published by McGraw Hill Book Company (p. 204).

The great German army was a shambles. On one of his first inspection tours of the P/W cages at a gigantic former compound of the Luftwaffe in Fuerstenfeldbruk near Munich, Patton had refrocked part of this unfrocked army. It was an SS cage, its inmates already segregated — presumably from the evils of their past. "On both sides of the broad road that ran through the camp," recorded the German Regiments-Adjutant, a Standarten-fuher only four weeks ago, arrested in the first wave of the round-up of SS officers, interned *sine die*, elected by his own comrades, and accredited to the *Amis*, "are housed two regiments of four battalions each, and a number of independent units. Every one of the battalions occupies its own cage, under joint command. The German command has its headquarters in one of the buildings on the airfield. Next door are the Americans."

The Prisoners, lined up to greet the American general, all had the typical earmarks of defeat — they were listless and empty eyed, clad in the tattered remnants of their uniforms. They were unshaven, unbathed, unkempt. Patton was appalled. He sent for German camp commandant, who turned out to be a former Wehrmacht colonel with the old snap. "Aren't you ashamed?" Patton yelled at him. "These men belonged to a great army only a few weeks ago. Look at them now! They are a bunch of vagrants at a convention of goddamn tramps. Get yourself together, Colonel. Show me that you know how a German soldier is supposed to look." Within twenty-four hours, the men were scrubbed and cleaned (Farago, Ladislas, p.204).

Farago, Ladislas. 1981.
The Last Days of Patton
Mc-graw-Hill Book Company, New York.
Material reproduced with permission of the publisher.

Waiting for the boat home

We were assigned to Heidelburg after our Fuerstenfeldbruk mission. Two weeks later we were assigned to support a Quartermaster Battalion working in Marseilles to expedite transfer of American troops to the Far-Eastern War. We encamped in French barracks north of Marseilles in Aix-en-Provence. Our battalion was slated for direct return to the United States, but all available ships were being used to transfer forces to the far east.

Most all American Army nurses, regardless of points accrued, were slated for the far east. They were encamped near Marseilles awaiting ships which would take them via the Suez Canal to the other war. It was anticipated that the forthcoming invasion of Japan would cost a million American casualties. Few of these American women were pleased about going to the Far East, but there were few complaints because these fine nurses were aware of the probable need for their services.

Lieutenant William Grohne, of Battery A, who had been a forward observer through most of the four campaigns, came down with Infantile Paralysis (POLIO), and was evacuated back to the states for years of treatment. Bill was a couple of years older than I. We had been together since early 1943. Our paths frequently crossed, including our stint in the church towers of Baesweiler. He was very courageous and exceptionally calm in combat. His troops were devoted to him.

V-J Day and demobilization

The word of the first atom bomb being dropped on Japan was glorious news. Troops could not be torn away from available radios. The second bomb created even greater enthusiasm. President Truman had prevented a million American casualties. The ensuing surrender of Japan created greater enthusiasm in the Marseilles region than most anywhere else in the world. Tens of thousands of American Army Nurses were released from assignment to the Asiatic front. The war was over and we could finally return to our families. Many of our officers had suffered a five-year family separation. The entire world was unified in a one-day super celebration. Someone from outer space would not have understood.

Our Army demobilized rapidly. In May 1945 there were 8,300,000 soldiers in the American Army. This was reduced to half by the end of 1945, and again by half by mid-1946. Eight million of our men made some use of the GI education benefits program.

Notable Facts about the Battalion

The 283rd Field Artillery Battalion served under three army groups, five armies, seven corps, eight divisions, and one cavalry squadron. This light artillery battalion's twelve changes in assignment were the result of being Theatre Artillery where units are sent to participate in major offensive drives. Citing from our *official Battalion History* (McDonald, 1945: page 76, Notable Facts):

> The 283rd Field Artillery Battalion was in combat for 268 continuous days...... The battalion travelled as a unit, a total

distance of 4241 miles in combat. The men from Highhedge (our radio code name) drove to the Rhine River with three different armies, Namely; the French First in Colmar on 8 February 1945, the American 1st Army in the Cologne area on 7 March 1945, and finally crossing with the American 7th Army just north of Worms, Germany on 26 March 1945. ....Highhedge howitzers fired a total of 45,361 rounds in combat. ... Some history making exploits follow: September 2, 1944 among the first American troops to enter Belgium and the second American artillery battalion in Belgium. ... 14 October to 5 November 1944 during this period while in position near Einighausen, Holland the Battalion, operating with the 125th Cavalry, held a 10,000 yard front and the left flank of all American Armies on the Western Front. November 21, 1944 to January 25th 1945.. were the most eastern of any allied artillery on the Western Front and the closest allied artillery to Berlin. On 26 January 1944 the Battalion left to take part in the Battle of Colmar. Travelling a distance of 320 miles with extremely adverse weather, over the Vosges Mountains during a blizzard, in 56 hours and 50 minutes, the longest march ever made by an artillery battalion in combat from one firing position to another. We were the first allied artillery to enter Colmar on 4 February 1945. March 6, 1945 the first artillery to cross the Roer River, and the first artillery to enter Cologne. April 17, 1945 the first artillery to enter Hitler's 'own city' Nuremberg, Germany. April 30, 1945 — in Munich, Germany. The town was entered from several sides. Highhedge was the first artillery unit to enter from the northeast (McDonald 1945).

The costs of the War in lives

The total costs of the war is far too large a topic for inclusion herein. The cost in lives was large but numbers don't tell the real story. Based on my own observation, of lost high school classmates, it is apparent many of our killed in action were the finest of their generation. Had these fine young Americans not given up their lives to quell the forces of evil they would have had equally fine descendants. Wars drain the best people from parenting equally fine offspring.

One of the achievements of our Battalion was the minimization of KIA (killed-in-action) losses to a total of only thirteen. Two of these were in C battery. The Battalion suffered numerous wounded, and four men were captured by the Germans. Our Battalion leadership sought to minimize casualties and they succeeded. A high casualty rate is not something to brag about. Our relatively few casualties, in what could have been disasters, were the result of the men continuously improving their defenses and other measures, including the use of common sense, judgment, and intelligence. The incident in Chapter 2 where Sergeant King was killed is a classic example of the use of superior judgement to avoid what could easily have become a firefight between our own men. The case at Colmar, where a 500-pounder landed in the middle of the battery and yet we had no casualties even warranting evacuation, is one example of the payoff from digging in. The direct-fire artillery duel at Nuremburg, where our KIA's were only one compared to the German fifty or more, is one more example. But probably, the best example was at Aldenhoven in November 1944 where "C" battery took in more than 500 German artillery rounds within the gun position without one KIA. Also, There were instances where our Battalion Commander and his staff, in meeting with the various artillery generals of the divisions to which we had just been assigned, discussed the role of our forward observers. I never knew exactly what transpired but I do know that we forward observers were treated royally by every unit which we worked with.

Considering the entire war and both theatres of operation, U.S. casualties were

Killed in action and died of wounds	= 237,049
Non-battle deaths	= 65,219

The total would be about 2/10ths of one percent of our population and less than half of our Civil War deaths without adjustment for our population then being one-fourth that of 1940.

German and Russian casualties were staggering. In 1943 it is estimated the Germans had 300 divisions of about 17,000 men (5.1 million), plus their Navy, Air Force, administrative/supply forces, and TODT (labor service). I estimate they had 8.5 million men under arms in 1943. Their total population was 79 million (including Austrians etc.). About 7 million German civilians were killed (mostly on the eastern regions) plus 3.25 million soldiers. The number of German POW's held by the Soviets in late 1945 were probably far more than one million. Reliable estimates are not available. Many died in captivity. In any event Germany lost 14.2 percent of their population and 37 percent of their losses were young men of one age class. Senior officers, quite willing to sacrifice teenagers, managed to protect themselves. Notice the contrast with Americans where the draft age remained 21 years until nearly one year after we entered the war. German men of my age have been scarce since the War. The few survivors of this age class were generally wounded, disabled, prisoners of the Russians until 1949, or a very few fortunate ones who were our prisoners.

Russian population was 210 million. Their deaths were approximately 12.9 percent or about 27 million (20 million soldiers along with 7 million civilians).

RETURN TO THE 1943 EXPERIMENT

Whereas no conclusions can be drawn on the experience of one battalion, it does seem that renewed experimentation should be considered. The combat performance of this one Battalion far exceeded expectations. The low casualty rate is astounding. Clearly we were a unified team. If the men in the battalion had undergone 17 weeks of basic training (Boot Camp) at one of the Replacement Training Centers the Battalion might still have performed as well, but I doubt the performance could have been better. Boot camp would have meant 17 fewer weeks in which the men were together building their sense of purpose. There was minimal bias in the selection of the men. They were average Pennsylvanians. Apparently, from available evidence, Pennsylvanians were, and are, typical Americans. The state does have the distinction of leading the nation with the highest percent of residents who were born in that state (80.2 percent). This facet probably has little bearing on the wartime performance of the one Battalion. It does explain why it was relatively easy to locate our veterans after a fifty-year hiatus. In 1994, at least seven were still living in their childhood homes.

Our gains from not having had "boot camp" included

1. Americans are practical inventive people who make wide use of expedients. Individualism, fortitude, restlessness, and exuberance highlight American characteristics (Turner 1920). Our men, never having been subdued by drill sergeants, retained their native exuberance.

2. Our men were average Pennsylvanians and, as far as I can determine, they were average Americans of that era. The induction people devoted a great deal of effort to selecting a cross-sectional average Pennsylvanian. However, once they were melted into this Battalion they were no longer average. It was tantamount to a chemical reaction. A catalyst for the "chemical reaction" was training sequencing coupled with leadership in the American style rather than the Prussian style.

3. Whereas discipline was a major problem in some RTC's and in many units (Palmer 1947), our men had the highest respect for all officers within the Battalion with the lone exception of the lieutenant who was transferred out while we were at Camp Rucker. Never once did I hear of any malicious or detrimental gossip about any of our officers. Even

fifty years later I am able to telephone all of our surviving veterans and be affably greeted. Widows of deceased veterans repeatedly cite past remarks regarding the fine officers we had. The phrase I have heard time and time again is

"I have been waiting fifty years to hear from you!"

4. Our officers didn't threaten the men with courts martial. There were only two courts martial in the Battalion during the two years, to my knowledge.

    There was the Fort Riley case where an officer, in some other unit, demanded a court martial for three of "A" Battery men. The men were departing the post exchange at Fort Riley when a second lieutenant was hurriedly entering and bumped into the three. He demanded an apology. Spunky Eugene E. Farra of Norristown, Pennsylvania, suggested it might be more appropriate if the lieutenant apologized to them. Hence "A" Battery's men were subjected to a court martial on the train departing Fort Riley for the Port of Embarkation.

5. Our men addressed officers by their title, seldom using the word "sir". I cannot explain the origin of this custom, except that it may have been more widespread throughout the Army than I knew about.

6. Our men had the highest respect for our non-commissioned officers although one first sergeant did seem to have some "run ins" with the men. In Chapter 4, I cited the case where one gun crew went to extremes to cover up the excess imbibing of alcohol by their sergeant.

    I have been receiving war-time photos from kinfolk of our veterans. A photo of one of our men was addressed to his sergeant with a caption that explains it all:

"To the best pal I ever had."

7. The exceptional respect for officers and non-commisnied officers, along with the exceptional comraderie, were partially attributed to our men having joined their units after only a few days in the Army well knowing that these were their comrades for the "duration".

There is little doubt that some people need boot camp prior to joining a military unit. However, the Army might be better off without this type of soldier. Possibly, boot camp requirements deter some highly intelligent youths from even joining the Army. And possibly, boot camps modify the attitudes of some highly intelligent soldiers in an adverse manner. Any new experiment on the sequencing of Army training might well screen out high intelligence recruits for direct assignment to operational units.

The ward-guardian relationship is established in any form of incarceration. Camp life, any sort of camp life a la longue, is itself wardship. It paralyzes and stultifies the personality. It is spiritual atrophy (Bailey 1991: p.149).

Our Battalion Commander

Lieutenant Colonel Hugh McDonald was a reserve officer. His performance, as battalion commander had been of the highest caliber. His objectives were to win the war, return to his family, and resume school teaching in St. Louis. The very little contact that I had with him was stimulating and pleasant. Obviously he was not concerned about his own promotion. Only recently, I realized that he was not popular with most officers of the battalion. Some officers and men have referred to him as a "loner". I have not located anyone who maintained any contact with him after the war. Perhaps "loners" make good commanders, I don't know. Lieutenant Colonel McDonald was highly intelligent, with a great amount of common sense. He evidenced his bravery on innumerable occasions. He allowed his officers a great amount of initiative as long as they were doing well. It seemed to me that this was our civilian army at its finest.

Veterans' benefits for our men and the Wehrmacht

Most officers and enlisted men of the 283rd Field Artillery Battalion were on active duty for three to five years. They placed their lives on the line and received only a pittance in pay. Their financial remuneration was but a tiny fraction of the wages of those who worked in war industry.

Our typical soldier was accorded a state bonus. Pennsylvania's was among the most liberal. However, we are speaking of a few hundred dollars. There also was the G.I. bill covering college education. Many of the veterans from this one battalion never enrolled under the G.I. Bill. Their anxiety to make up lost time by getting married and rearing a family seems to have obviated college. Their offspring, however, have had superb education, good family discipline, have a most favorable attitude, and are our hope for the future.

The G.I. Bill was an excellent program in two aspects. First, it vastly furthered America's human resource potential. Secondly, it removed millions of young men from the job market at a time when economic forecasts predicted a recession. There was no federal bonus such as that for the World War I veterans, which was as much as $1,500. When this was paid in 1936 it was enough to make full payment on a home. There was one Congressional proposal to open up Alaskan lands to veteran settlement. It was similar to other programs following most all of America's wars. Frank Heintzleman, the Territorial Governor of Alaska, persuaded Congress to shelve that proposal. There is a small burial allowance. And Veteran's Administration hospitals are open to veterans on a space available basis. Those with combat disabilities have a priority for hospital admission.

In the 1990's when their limited benefits are systematically being reduced it is appropriate to compare U.S. war-time veteran benefits with those of the Wehrmacht and the Waffen SS.

Recently, the TV program "Sixty Minutes" featured former SS Latvians who were drawing pensions from Germany. Those men from other countries who had joined the German army were promised all of the rights of a German soldier. However, this TV program may not be the most reliable source of information. James M. Diehl, in his 1993 book *The Thanks of the Fatherland* reports that from the end of the war until 1951 German veterans, excepting disabled and former prisoners of war, received no pensions nor other benefits. However, since 1951 German career officers have been receiving pensions. Noncommissioned officers have not received pensions although there have been other benefits for some.

Survivors in 1994

In 1994, 44 percent of the 219 men and officers from our battalion whom I have searched out were deceased. This is about average survival. *Life Tables for the United States 1900-2050*, published by the U.S. Department of Health and Human Services, suggests that fifty percent of the men, who were born in 1922 should have lived until the year 1993. The extreme stress of 268 continuous days of combat in the European Theater apparently did not effect future health. The kinfolk of our KIA and deceased are most important to us. We have established contact with many of them and are striving to reach out to all widows and descendants. We have also set up a procedure for enabling descendants to reach out to each other, and to procure copies of the original unit history for at least the next century. There will always be an occasional youth who wants to know more about his/her great grandfather.

I was released from active duty as a Captain in September 1945 at Indiantown Gap, Pennsylvania. When I arrived at the North Philadelphia Train Station it developed that all taxi drivers were providing free transport home to returning veterans. When we arrived home the taxi driver dallied long enough to view the homecoming which was his reward.

I elected to stay in the Reserves as a captain while I attained my BS in Forestry degree at West Virginia University and my Master Forester degree from Yale University. In 1949 I accepted a Regular Army Commission, as a first lieutenant, in the Corps of Engineers. Unknowingly, I was becoming involved again, this time in the Korean War. Six campaign stars on my Korean War ribbon represent the six campaigns where I was in front-line combat for fourteen months with the 8th Engineer Battalion of the First Cavalry Division. Our initial forces in Korea were Regular Army. A new book is underway.

Map 1. They came from 102 Pennsylvania towns and cities

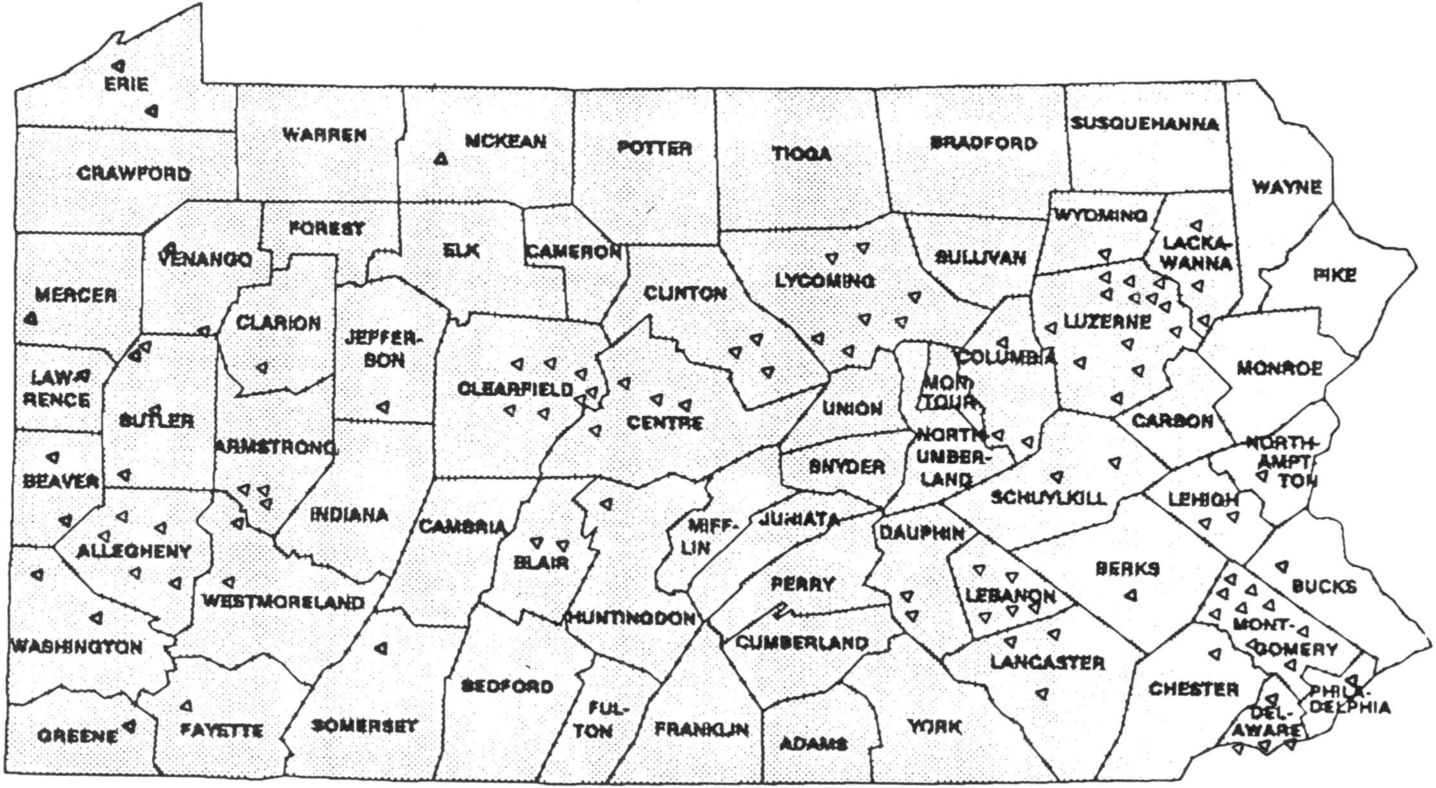

163

Map 2.

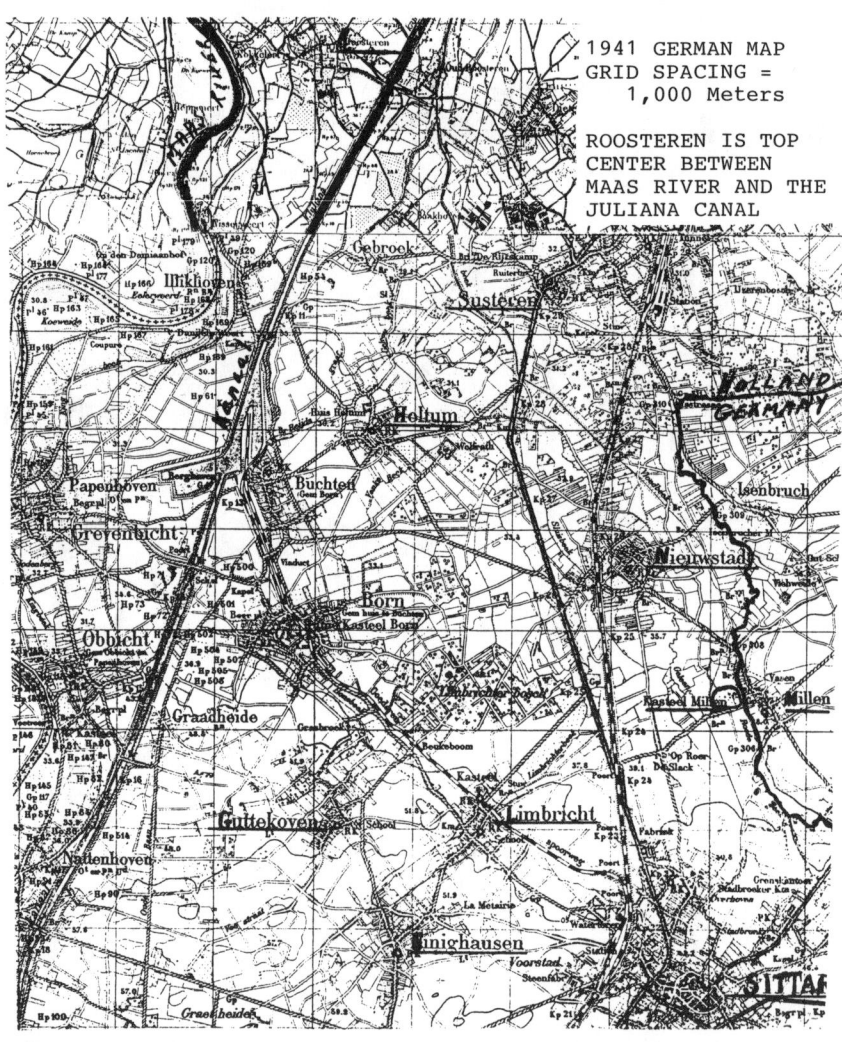

1941 GERMAN MAP
GRID SPACING =
1,000 Meters

ROOSTEREN IS TOP
CENTER BETWEEN
MAAS RIVER AND THE
JULIANA CANAL

165

Map 3.

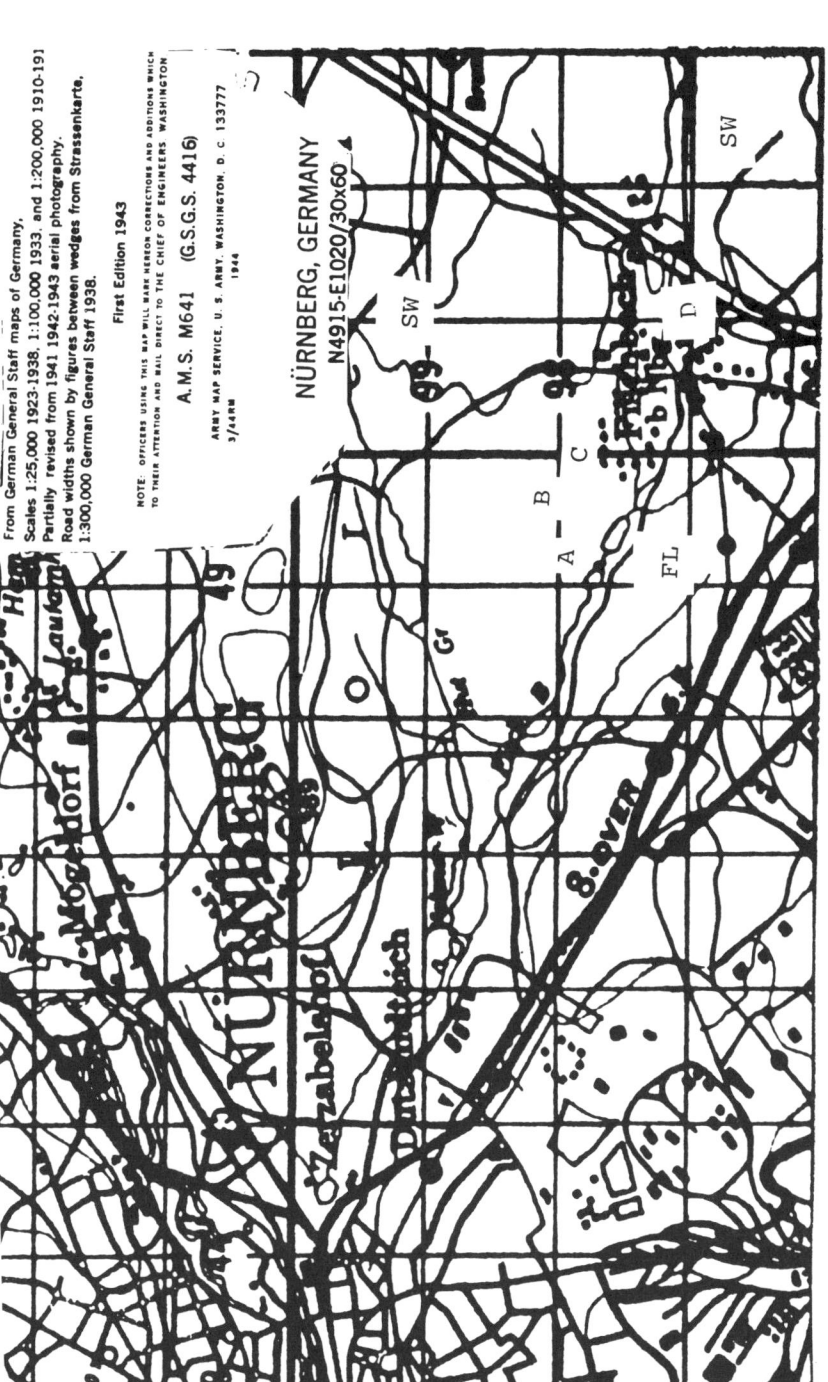

Map 4.

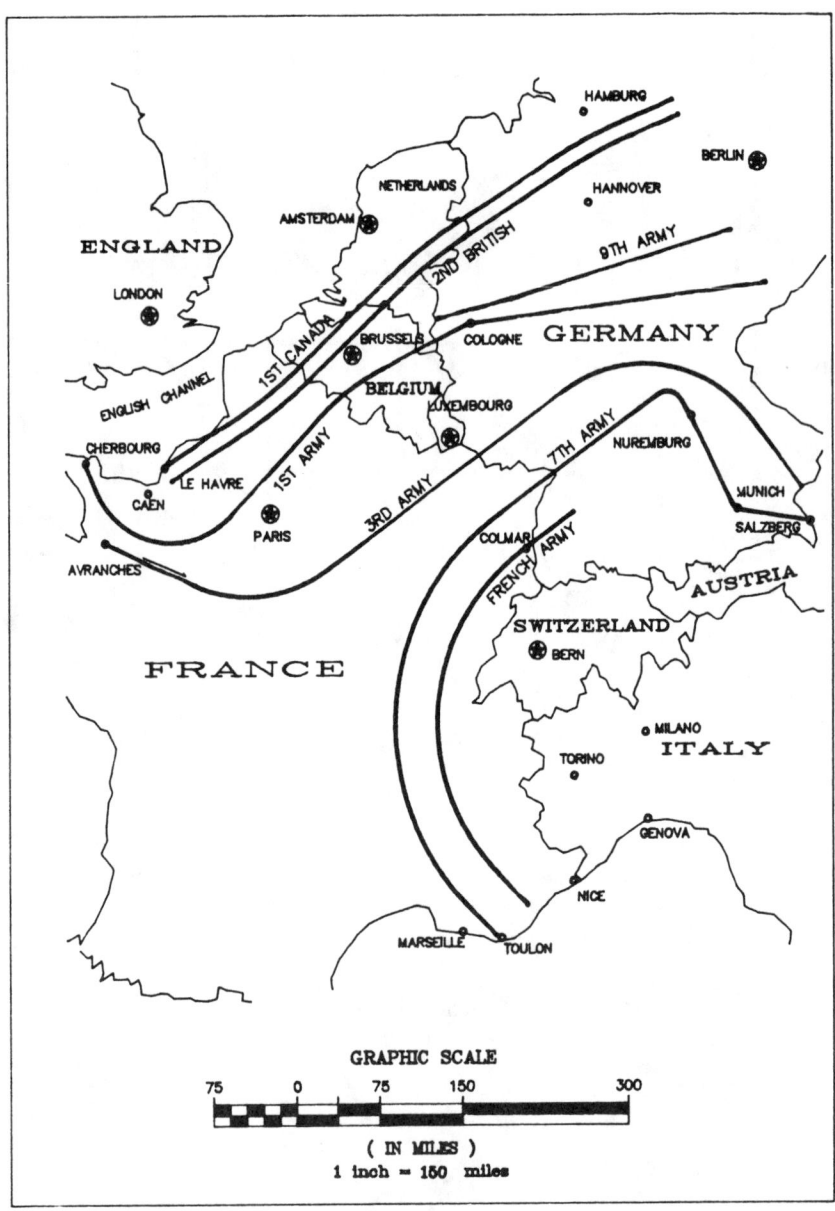

GRAPHIC SCALE

75   0   75   150   300

(IN MILES)
1 inch = 150 miles

Map 5.

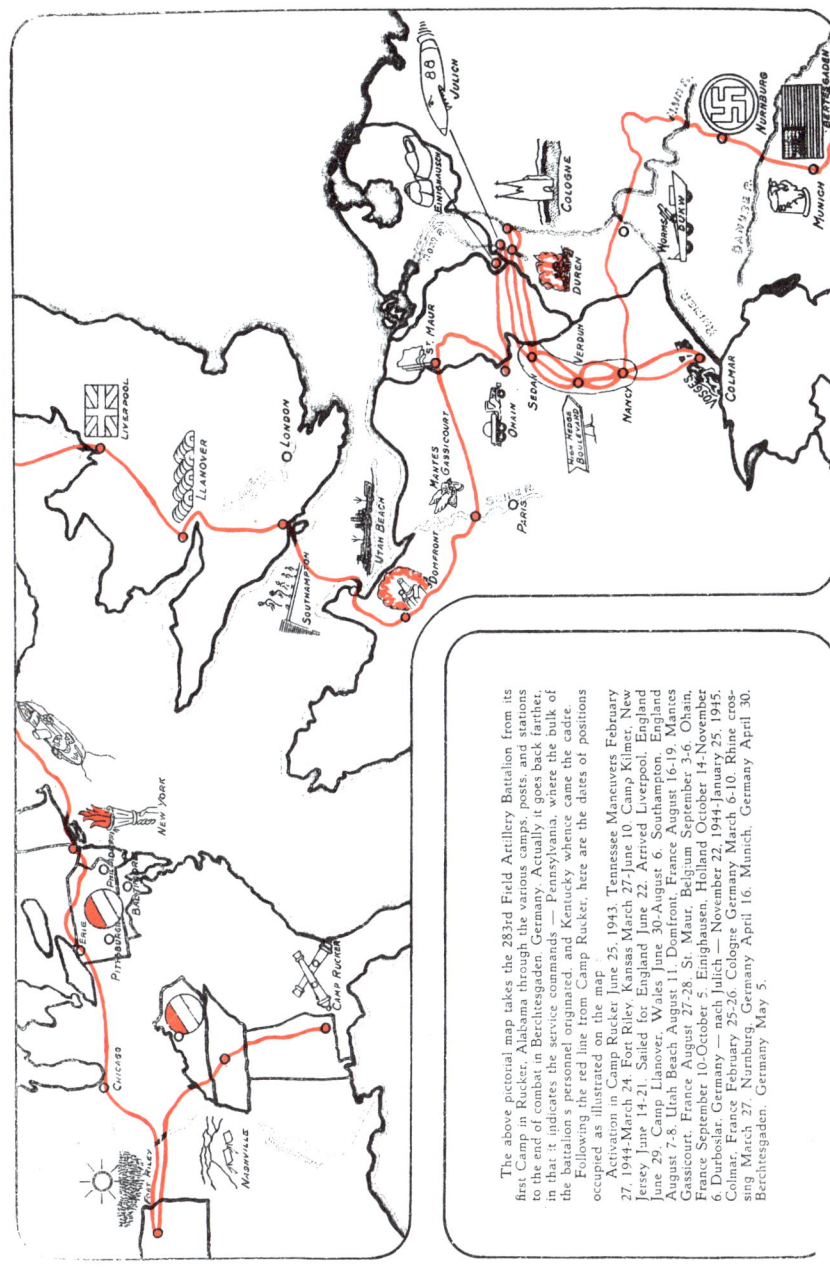

The above pictorial map takes the 283rd Field Artillery Battalion from its first Camp in Rucker, Alabama through the various camps, posts, and stations to the end of combat in Berchtesgaden, Germany. Actually it goes back farther, in that it indicates the service commands — Pennsylvania, where the bulk of the battalion's personnel originated, and Kentucky whence came the cadre.

Following the red line from Camp Rucker, here are the dates of positions occupied as illustrated on the map:

Activation in Camp Rucker June 25, 1943. Tennessee Maneuvers February 27, 1944-March 24. Fort Riley, Kansas March 27-June 10. Camp Kilmer, New Jersey June 14-21. Sailed for England June 22. Arrived Liverpool, England June 29. Camp Llanover, Wales June 30-August 6. Southampton, England August 7-8. Utah Beach August 11. Domfront, France August 16-19. Mantes Gassicourt, France August 27-28. St. Maur, Belgium September 3-6. Ohain, France September 10-October 5. Einighausen, Holland October 14-November 6. Durboslar, Germany — nach Julich—November 22, 1944-January 25, 1945. Colmar, France February 25-26. Cologne Germany March 6-10. Rhine crossing March 27. Nurnburg, Germany April 16. Munich, Germany April 30. Berchtesgaden, Germany May 5.

Itinerary of 283rd Field Artillery Battalion

DATE   BATTALION HQ.   C BATTERY

Campaign for Northern France

12th U.S. Army Group; 3rd U.S. Army; XII Corps

9 Aug 1944	Utah Beachhead, France	Utah Beachead, France
11 Aug 44	12th Corps Marshalling Area	8 Mi to 12th Corps Marshalling
12 Aug 44		35 Mi to Vic. Cherbourg, France

12th U.S. Army Group; 1st U.S. Army; XIX Corps; 30th Inf. Div.

13 Aug 44	Millny, France	80 Mi to Millny, France
14 Aug 44	St. Mars, France	9 Mi to St. Mars, France
15 Aug 44		
16 Aug 44	N.E. Domfront, France	10 Mi to Domfront, France
17 Aug 44		
18 Aug 44		
19 Aug 44		
20 Aug 44	Brezalles, France	109 Mi to Crucey, France
21 Aug 44	Le Souchet, France	18 Mi to Le Souchet, France
22 Aug 44	St. Luc, France	

12th U.S. Army Group, 1st U.S. Army; XV Corps; 30th Inf. Div.

23 Aug 44		12 Mi to Evreux, France
24 Aug 44		
25 Aug 44	Emmalville, France	13 Mi to Verdun, France (Nord)
26 Aug 44	Verdun, France	
27 Aug 44	Lemay, France	25 Mi to Mantes-Gassicourt, Fr.
		Crossed Seine River
28 Aug 44		

12th U.S. Army Group; 1st U.S. Army; XIX Corps; 30th Inf. Div.

29 Aug 44	Lainville, France	
30 Aug 44	Gericourt, France	28 Mi to Gennicourt, France
31 Aug 44	Thermicourt, France	18 Mi to Neuilly, France
1 Sep 44	Enroute	125 Mi to
2 Sep 44	Enroute	St. Maur, Belgium

3 Sep 44	St. Maur, Belgium	
4 Sep 44		
5 Sep 44		
6 Sep 44		4 Mi to Fontency, Belgium
7 Sep 44	Gaillernarde, Belgium	60 Mi to Goillemorde, Belgium
8 Sep 44		
9 Sep 44	Gaillernarde, Belgium	
10 Sep 44	Ohain, France	65 Mi to Ohain, France
until	Red Ball Express	
15 Sep 44		
16 Sep 44	Ohain, France	

Campaign for the Rhineland

12th U.S. Army Group; 9th U.S. Army; XIX Corps; 103rd Cavalry Group

5 Oct 44	Ohain, France	123 Mi to Teuvan, Belgium
6 Oct 44	Teuven, Belgium	
until		
13 Oct 44	Teuven, Belgium	29 Mi to Guttegowen, Holland
14 Oct 44	Einighausen, Holland	
until		
6 Nov 44	Einighausen, Holland	18 Mi to Boutsch, Holland
		Close by Heerlen, Holland

12th U. S. Army Group; 9th U.S. Army; XIX Corps; 29th Inf. Div.

7 Nov 44	Bautsch, Holland	10 Mi to Merkstein, Germany
8 Nov 44	Floes, Germany	
until		
20 Nov 44	Floes, Germany	6 Mi to Bettendorf, Germany
21 Nov 44	Bettendorf, Germany	4 Mi to Aldenhoven, Germany
22 Nov 44	Duboslar, Germany	
until		
20 Dec 1944		

21st British Army Group; 9th U.S. Army; XIII Corps; 29th Inf. Div.

25 Jan 1945	Duboslar, Germany

6th U.S. Army Group; First French Army; XXI Corps; 28th Inf. Div.

26 Jan 45   Verdun, France. Enroute   179 Mi to Verdun, France

Campaign for Alsace-Lorraine

27 Jan 45   St. Die, France. Enroute   138 Mi to St. Die, France

28 Jan 45   35 Mi to Ostheim, France

29 Jan 45   Ostheim, France

30 Jan 45

31 Jan 45

 1 Feb 45

 2 Feb 45

 3 Feb 45   Colmar, France   7 Mi to Colmar, Alsace Lorr.

 4 Feb 45

 5 Feb 45   3 Mi to E. Colmar, Alsace Lor.

 6 Feb 45   South Colmar, France   6 Mi to Niederbergheim,
                                                      Alsace Lor.

 7 Feb 45   Niederbergheim, France   3 Mi to Dessenheim, Alsace L.

 8 Feb 45   Dessenheim, France

 9 Feb 45

10 Feb 45

11 Feb 45   80 Mi to Longechamp, France

12 Feb 45   Longechamps, France

13 Feb 45

14 Feb 45

15 Feb 45

16 Feb 45

17 Feb 45

12th U.S. Army Group; 1st U.S. Army; VII Corps; 104th Inf. Div.

18 Feb 45   Enroute

19 Feb 45   178 Mi to Sedan, France

Continuation of Campaign for the Rhineland

20 Feb 45   120 Mi to Vergath, Germany
                                                      Vic. of Eschweiler

21 Feb 45   Eschweiler, Germany

22 Feb 45		5 Mi to Luchen, Germany on west bank of Roer across from Duren, Germany
23 Feb 45		
24 Feb 45	Luchem, Germany	3 Mi to Gurzenich, Germany
25 Feb 45	Gurzenich, Germany	
26 Feb 45		6 Mi to Merzenich, Germany
27 Feb 45	Merzenich, Germany	
28 Feb 45	Gut HS Forst, Germany	5 Mi to Mannheim, Germany
1 Mar 45		
2 Mar 45	Sindorf, Germany	7 Mi to Sinsdorf, Germany
3 Mar 45	Konigsdorf, Germany	
4 Mar 45		6 Mi to Konigsdorf, Germany
5 Mar 45		
6 Mar 45	Weiden, Germany	2 Mi to Weiden, Germany 3 Mi to Cologne, Germany
7 Mar 45	Cologne, Germany	
8 Mar 45		
9 Mar 45		

6th U.S. Army Group; 7th U.S. Army; XV Corps; 44th Inf. Div.

10 Mar 45	Verdun, France. Enroute	190 Mi to Verdun, France
11 Mar 45	Guising, France	100 Mi to Guising, France
12 Mar 45		

6th U.S. Army Group; 7th U.S. Army; XV Corps; 45th Inf. Div.

13 Mar 45	Saareinsming, France	10 Mi to Saareinsming, France
14 Mar 45		
15 Mar 45	Blies River Crossing	
16 Mar 45	Habkirchem, Germany	8 Mi to Habkirchen, Germany 3 Mi to Wittersheim, Germany
17 Mar 45	Wittersheim, Germany	
18 Mar 45	Ballweiler, Germany	4 Mi to Ballweiler, Germany
19 Mar 45		
20 Mar 45		
21 Mar 45		15 Mi to Beeden (by Homburg)
22 Mar 45	Beeden, Germany	50 Mi to Heidersheim, Germany

BATTLE OF CENTRAL EUROPE

(Battle of Inner Germany)

23 Mar 45	Heidensheim, Germany	
24 Mar	Herrnsheim	11 Mi to Herrensheim, Germany
25 Mar		
26 Mar		
27 Mar	Rhein-Durkheim	6 Mi to RheinDurkheim, Germany
		13 Mi to Felheim, Germany
		20 Mi to Reichenbach, Germany
28 Mar	Reichback	12 Mi to Grbeiberau, Germany
		7 Mi to Oberklingen, Germany
		10 Mi to Neustadt (Main)
29 Mar	Monrlingen	4 Mi to Monlingen, Germany
30 Mar	Hofstetten	14 Mi to Elsenfeld, Germany
		2 Mi to Hofstetten, Germany
31 Mar		
1 Apr	Soden	8 Mi to Soden, Germany
2 Apr		26 Mi to Schollkrippen, Germany
3 April	Obndorf	18 Mi to Oberndorf, Germany
4 April	Jossa	10 Mi to Jossa, Germany
5 Apr		17 Mi to Ober Kalbac, Germany
6 Apr	Siefert	27 Mi to Wickers, Germany
		5 mi to Sieforts, Germany
7 Apr		
8 Apr	Lothausen	44 Mi to Lothausen, Germany
9 Apr		
10 Apr		
11 Apr	Lofsbergsgereuth	20 Mi to Lofsbergsgereuthe, Ger.
12 Apr	Unter Oberndorf	3 Mi to Rattlesdorf, Germany
13 Apr	Gurdelsheim	5 Mi to Unter Oberndorf, Germany
		5 Mi to Gundelsheim, Germany
14 Apr		
15 Apr		
16 Apr	Fischbach	55 Mi to Fischbach, Germany
17 Apr		
18 Apr		
19 Apr		

20 Apr		
21 Apr		
22 Apr	Georgesmund	42 Mi to Georgesmund, Germany
23 Apr	Kotterhochstadt	20 Mi to Kattenhochstadt, Ger.
24 Apr	Rahlingen	17 Mi to Rahlingen, Germany
25 Apr	Kierberg	16 Mi to Amersfeld, Germany
26 Apr		
27 Apr	Etting	18 Mi to Etting, Germany
28 Apr	Glonn	38 Mi to Gloon, Germany
29 Apr	Ampermoching	9 Mi to Ampermoching , Ger.
30 Apr	Feldmoching	9 Mi to Munich, Germany

6th U.S. Army Group; 7th U.S. Army; XV Corps; 42nd Inf. Div.

1 May	Perlach	11 Mi to Perlach, Germany
2 May	Esfernef	8 Mi to Essernef, Germany
3 May	Rottous	55 Mi to Grassau, Germany

6th U.S. Army Group; 7th U.S. Army; XXI Corps; 3rd Inf. Div.

4 May	Mouthsn	42 Mi to Mauthen, Germany Salzburg, Austria Berchtesgaden, Germany
5 May	Berchtesgaden	
6 May		
7 May	END OF WAR	
8 May		
9 May		71 Mi to Munich, Germany
10 May	Munich	
11 May		
12 May		
13 May		
14 May	Furstenfeldbruck	21 Mi to Furstenfeldbruck, Ger.
Summary:	3 army groups	
	5 armies	
	7 corps	
	8 divisions	
	1 Cav. Sq.	

Order of Battle – The First 56 Days

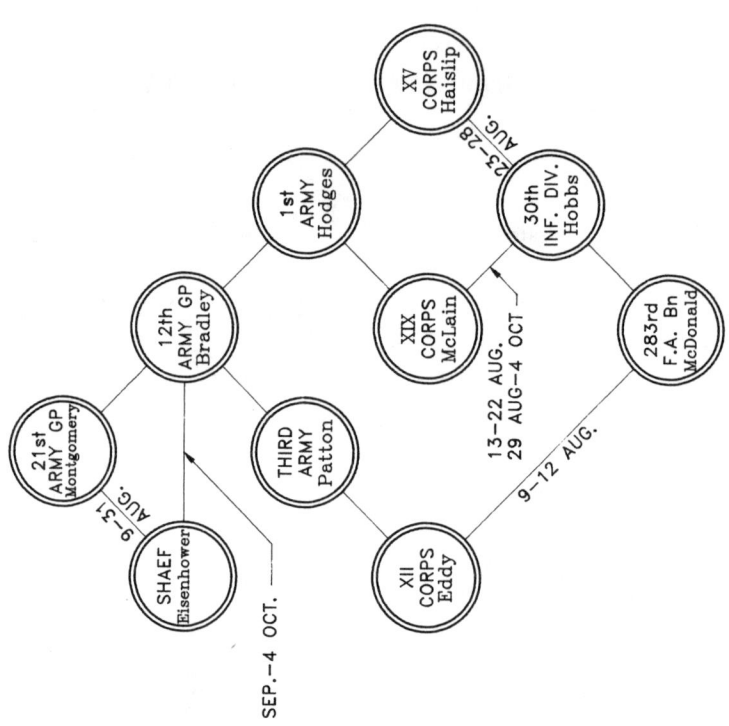

Order of Battle — The Next 76 Days

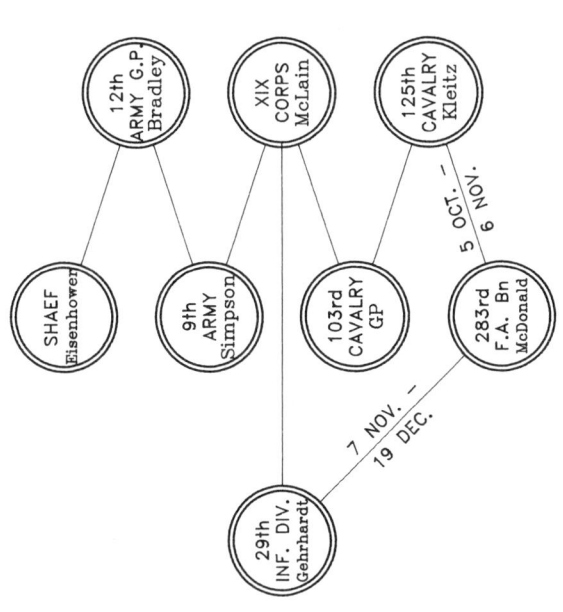

Order of Battle — The 37 Days of the Ardennes Offensive

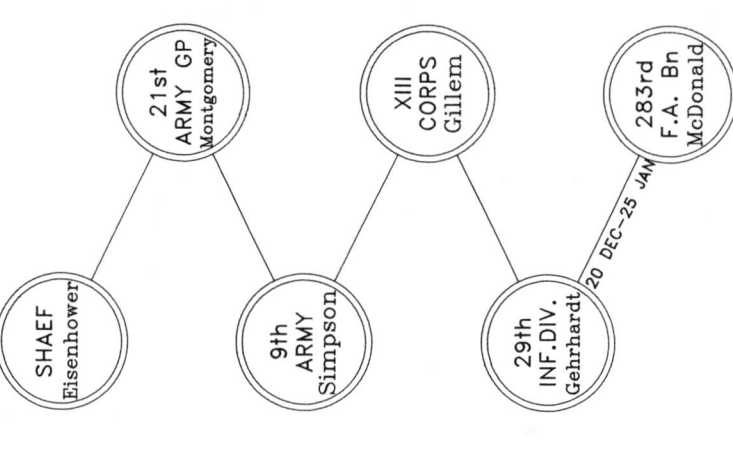

Order of Battle – The Final 102 Days

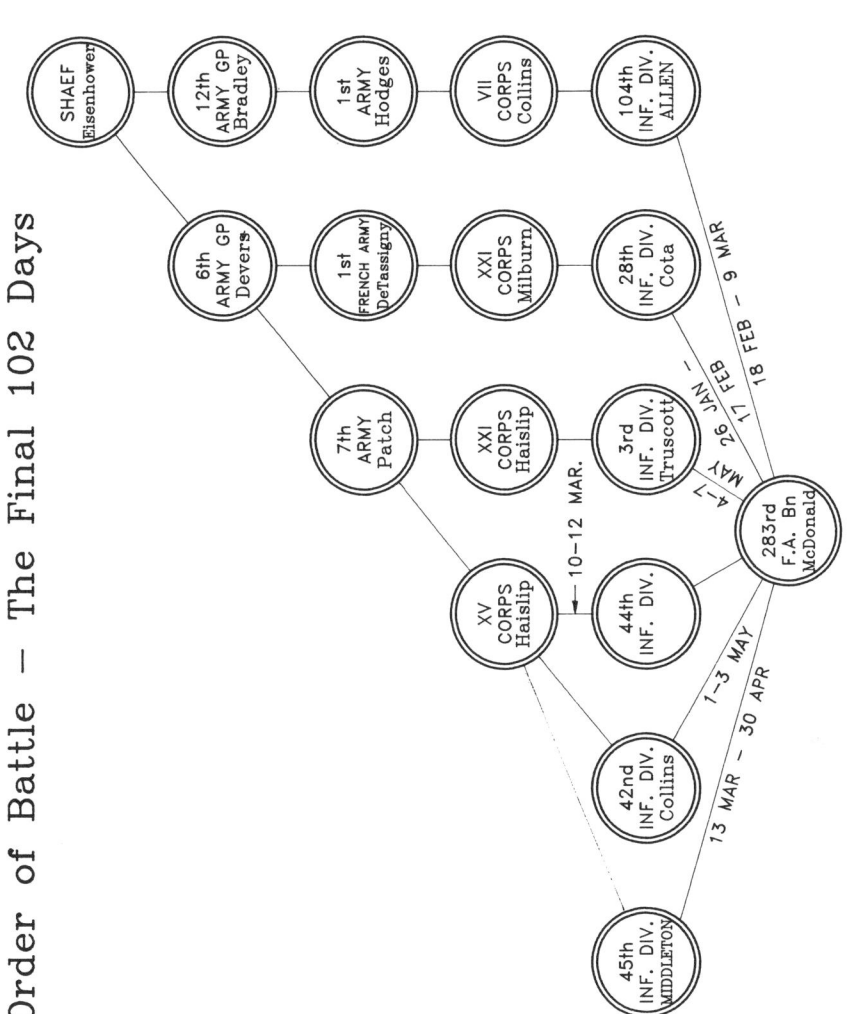

Appendix C
United States' Order of Battle
with Artillery Emphasis

Information compiled from and reprinted with permission from *Order of Battle World War II* by Shelby L. Stanton c1984

Published by Presidio Press, 505 B San Marin Drive, Novato, CA 94945

NOTE: This compilation pertains to the European Theatre of Operations (ETO) excluding the Mediterranean Theatre, unless otherwise noted.

	U.S. Total	European Theatre
Armies	11	6
Corps	26	14
Divisions, Infantry	71	41
Divisions, Airborne	5	4
Divisions, Armored	16	16
Artillery Group Hq.	102	62
Artillery Obsvn. Bns.	21	16

The arrival of sixty-one U.S. divisions in France.

	Infantry	Armored	Airborne	Total
June	1 2 4 9 29 30 79 83 90	2 3	82 101	13
July	5 8 28 35	4 5 6		20
Aug.	3* 36* 45* 80	7		25
Sept.	26 44 94 95 102 104	10		32
Oct.	100 103	9 14		36
Nov.	78 84 99	12		40
Dec.	66 75 87 106	11 17		46
Jan.	42 63 65 69 70 76 89	8 13		55
Feb.	71	16 20	13	59
Mar.	86 97			61

* landed in southern France

The fourteen corps in the European Theatre of Operations (ETO).

III	V	VI	VII	VIII	XII	XIII	XV	XVI
XIX	XX	XXI	XXII	XXIII	and XVIII Abn			

The sixty-four artillery group headquarters units in ETO.

5	6	17	18	30	35	36	40	46
79	119	142	144	153	173	174	177	179
182	183	187	188	190	193	194	195	196
202	203	204	205	208	209	210	211	212
214	219	220	224	228	250	252	258	333
349	351	401	402	404	405	406	407	408
410	411	413	416	417	421	422	425	426
472								

The four hundred and eighty-four artillery battalions in ETO.

Divisional Artillery	224 battalions
Non-Division Artillery	
75mm Prcht Bn	3 battalions
Armored 105 mm	16 battalions
Towed 105 mm howitzers	37 battalions
155 mm Howitzers	72 battalions
4.5-inch gun	17 battalions
155mm Gun	38 battalions
8-inch Howitzers	37 battalions
8-inch Gun	5 battalions
240 mm Howitzers	15 battalions
Captured enemy wpns.	1 battalions
Field Arty. Obsn. Bn.	19 battalions

Total battalions = 484 (of which 260 were non-divisonal)

Non-Divisional battalions numbered designations

75mm Parachute Field Artillery Battalions

464	601	602	= 3 battalions

105mm Armored Battalions =

58	59	62	65	69	83	87	93	253
274	275	276	342	400	695	696	= 16 Bns.	

105mm Towed Battalions =

18	25	70	74	76	115	130	162	170
193	196	241	242	250	252	255	280	281
282	283	284	394	401	402	512	522	569
580	583	627	687	688	690	691	692	693
802	= 37 battalions							

155mm Howitzer Battalions

2	17	30	36	81	141	177	179	182
183	186	187	188	191	202	203	204	209
215	228	254	257	333	349	350	351	521
550	665	666	667	670	671	672	673	686
689	751	752	753	754	755	758	759	761
762	763	764	767	768	805	808	809	933
937	938	940	942	943	945	949	951	953
955	957	961	963	965	967	969	974	975

= 72 battalions

155mm Gun Battalions

77	174	190	200	208	240	258	261	273
514	515	516	528	540	541	546	547	548
549	557	558	559	561	634	635	731	733
734	976	977	978	979	980	981	987	989
991	993	= 38 battalions						

4.5-inch Gun Battalions

172	176	199	211	259	770	771	772	773
774	775	776	777	935	939	941	959	

= 17 battalions

8-inch Howitzer Battalion

194	195	207	264	529	535	578	630	656
657	658	659	660	661	662	663	736	738
739	740	741	742	743	744	745	746	747
748	787	788	790	791	793	932	995	997
999	= 37 battalions							

8-inch Gun Battalions

153	243	256	268	575	= 5 battalions

240mm Howitzer Battalions

265	266	267	269	270	272	277	278	538
539	551	552	553	697	698	= 15 battalions		

Battalions with captured enemy weapons = 244th F. A. Bn.

Field Artillery Observation Battalions

1	2	3	7	8	12	13	14	15
16	17	285	286	288	290	291	292	293
294	= 19 battalions							

Towed 105mm non-divisional howitzer battalions organization/training

   17 old Army from former regiments

    7 organized in Feb. 1943

    2 organized in May 1943

    4 organized in June 1943

    6 organized in Apr. 1944

The May 1943 battalions were the 280 and the 281 at Camp Cooke.

The June 1943 were the 282 283 284 512 at Camp Rucker.

The 283rd was the first of this group to land in France

Towed 105 mm non-divisional howitzer battalions arriving in France.

   June 1944 =   0

   July 1944 =   7

   Aug. 1944 = 11 (283rd was the first of these)

   Sept. 1944 =   5

   Oct. 1944 =   2

   Nov. 1944 =   1

          1945 = 11

Artillery men, warrant officers, and officers in the E.T.O. at War's end.

14 Corps Headquarters	×	116	personnel	=	1,624
62 Artillery Groups	×	99	personnel	=	6,138
42 Infantry Divisions	×	2,160	personnel	=	90,720
15 Armored Divsions	×	1,623	personnel	=	24,345
4 Airborne Divisions	×	1,424	personnel	=	5,696
16 Armored 105mm Bns.	×	545	personnel	=	8,720
37 Towed 105mm Bns.	×	521	personnel	=	19,277
72 155mm Howitzer Bns.	×	531	personnel	=	38,232
17 4.5-inch Gun Bns.	×	531	personnel	=	9,027
38 155mm Gun Bns.	×	506	personnel	=	19,228
37 8-inch Howitzer Bns.	×	581	personnel	=	21,497
5 8-inch Gun Bns.	×	490	personnel	=	2,450
15 240mm Howitzer Bns.	×	490	personnel	=	7,350
1 Enemy Weapons Bn.	×	500 (est.)	personnel	=	500
19 Artillery Obsvn Bn.	×	466	personnel	=	8,854
			Total	=	263,658

Estimated seven percent non-returning casualties incl. KIA.

263,658 × 1.07 = 282,114 artillery men, warrant officers, and officers in the European Theatre out of a total three million (approx.). This is 9.4 percent.

Bibliography/Literature Cited

Ambrose, Stephen E. 1981. *Ike's Spies — Eisenhower and the intelligence community* (Story of the CIA), Doubleday, NY

Anderson, Milton (former sergeant in Service Btry) Disc. 1993. Lancaster, Pennsylvania.

Anthony, Evelyn. 1986. *Voices in the Wind* G. Putnam & Sons.

Arkin, William M. et al. 1990. *Encyclopedia of the U.S. Military* Harper & Row

Armstrong, Frank H. and Marguerite O. Oates. 1992. *Window Seat — An Aerial Perspective of America's Forests* Bull Run of Vermont, Inc.

Baender, Paul. 1992. *A hero perished — The diary and selected letters of Nile Kinnick*. (The story of WWII pilot Nile Kinnick, the Iowa State Heisman Trophy Winner) University of Iowa Press.

Bailey, George. 1991. *The Biography of an Obsession — Germans* The Free Press, NY.

Baldwin, Hanson W. 1979. *Tiger Jack* (Major General John S. Wood and the 4th Armored Division in Europe). Introduction by General Jacob L Devers. Old Army Press.

Balfour, Michael. 1982. *West Germany, a contemporary history* St. Martin's Press.

Barnes, Kenneth C. 1991. *Nazism, Liberalism, & Christianity Protestant social thought in Germany and Great Britain 1925-1937* The University Press of Kentucky

Bartov, Omer. 1991. *Hitler's Army — Soldiers, Nazis, and War in the Third Reich* Oxford University Press, New York, Oxford.

Bassett, Richard. 1989. *Waldheim and Austria* Viking Press. NY

Beck, Philip. 1979. *Oradour — Village of the Dead* Leo Cooper, London.

Beebe. Gilbert W., Ph.D., and Michael E. De Bakey, M.D. 1952. *Battle Casualties — Incidence, Mortality, and Logistic Considerations* Charles Thomas — Publisher, Springfield, IL

Bennett, Ralph Francis. 1980. *Ultra in the West: The Normandy Campaign 1944-1945.* Scribner, New York.

van den Berk, H Th G. 1993. Minnikstraat 24, NL 6123 AR Holtum, Netherlands. Personal correspondence 1993 series.

Bischof, Gunter and Stephen E. Ambrose. 1993. *Eisenhower and the German POWs* Louisiana State University Press.

Blumenson, Martin. 1963. *The Duel for France 1944* Houghton-Mifflin

Blumenson, Martin. 1961 *Breakout and Pursuit*

Boas, Jacob. 1985. *Boulevard des Miseres — the story of transit camp Westerbork in WWII* Archon Books.

Botting, Douglas. 1983. *The Aftermath — Europe* Time-Life Books

Boyd, Carl. *Hitler's Japanese Confidant — General Oshima Hiroshi and Magic Intelligence 1941-1945* 1993. University Press of Kansas, Lawrence.

Bradford, Ernle. 1986 . *Siege Malta 1940-1943* William Morrow and Company, Inc., New York.

Brands *The Federal Republic of Germany and the Netherlands* Brenner, Myron (former sergeant in "C" Btry) Corresp. & Disc 1993/1994. Hazleton, Pennsylvania.

Briels, Jack. 1993. Personal correspondence and pages from a draft of his autobiography. 5665 South 1475 E, #5D, South Ogden, Utah 84403 Tel 801 479 5002. Telecons and Correspondence 1993 & 1994.

Bryant, R.C. 1936. Wood gas as a motor fuel. Journal of Forestry Vol. 34, No. 8:816-817.

Buechner, Emajean. 1991. *Sparks — The combat diary of a battalion commander (rifle) WWII* Thunderbird Press, Metairie, LA

Bullen, R. J. 1984. *Ideas into Politics — Aspects of European History 1880-1950* Barnes and Noble, London.

Burns, James MacGregor. 1970. *Roosevelt: The Soldier of Freedom 1940-1945* The concluding volume of the first complete biography of FDR. Harcourt Brace Jovanovich, New York.

Butler, Ewan (Lt. Col.) and Selby Bradford (Major). 1950. *The Story of Dunkirk* Hutchinson, London.

Burt, Kendal and James Leasor. 1956. *The One that got Away* (Escaped German pilot). Random House.

Cairncross, Alec. 1986. *The Price of War — British policy on German reparations 1941-1949* The author was chief negotiator. Blackwell, New York.

Cawthon, Charles R. 1990. *Other Clay — a remembrance of the World War II Infantry.* (29th Division). University Press of Colorado.

Collins, Lawton J., General. 1945. *Mission Accomplished — The Story of the Campaigns of the VII Corps United States Army in the War Against Germany 1944-1945* J.J. Weber, Leipzig.

Compton, Wilson. 1942. National Lumber Manufacturers Report Journal of Forestry Vol. 40, No. 2: 914

Cowburn, Benjamin. 1960 *No Cloak, No Dagger* (Behind the lines in France 1941-1944). Jarrolds, London.

Craig, Gordon A. 1982 *The Germans* Putnam & Sons.

Davis, William (former battalion personnel clerk). Corresp. 1994. St. Charles, Missouri.

Deane, Drummond Anthony. 1954 *Return Ticket* (Demolition behind enemy lines in Italy). J. B. Lippincott Co.

Demetz, Hana. 1980. *House on Prague Street* (The story of a half-Jew in Prague during the war. She was never placed in an internment camp) St Martin's Press, New York.

Denham, M. 1985. *Inside the Nazi Ring (A naval attache in Sweden, 1940-1945* Holmes & Meier, New York.

Denton, James (former officer in "B" Btry) corresp. & Disc. 1993/1994 Short Hills, New Jersey.

Department of the Army. 1959. *American Military History 1607-1958* ROTCM 145-20. Headquarters, Department of the Army.

Derry, Sam (Lt. Col.) 1960. *The Rome Escape Line* (The British organization in Rome for assisting escaped prisoners of war.) W.W. Norton Company, New York

Devigny, Andre. 1951. *Escape from Montluc* Dennis Dobson, London.

Devitz, Carl. 1993. 714 N. 3rd Ave., Lebanon, PA 17042 (717) 272 8409. Personal telecons.

Diehl, James M. 1993. *The Thanks of the Fatherland — German Veterans After the Second World War* Univ. of North Carolina Press, Chapel Hill and London.

Dieterle, Frederick (former Battalion Exec.) 1993/1994. Correspondence. Santurce, Puerto Rico.

Dornberg, John. 1982. *Munich 1923 — The story of Hitler's Putsch* (excellent listing of the characters and their fate at the end of the book) Harper and Row, New York.

Dotterer, Eugene. 1994. Personal correspondence & Telecon.

Duffy, Christopher. 1991. *Red Storm on the Reich* Atheneum Press.

Eagen, William. 1975 *The Man on the Red Horse* Unit history of the 113th Cavalry Group. Metropolitan Printing Co., Portland, OR.

Egger, Bruce E and Lee MacMillan Otts. 1992. *"G" Company's War — two personal accounts of the campaign in Europe 1944-1945.* University of Alabama Press.

Elkins, Stanley and Eric McKitrick. 1954. A meaning for Turner's frontier Part I, Democracy in the old Northwest, Political Science Quarterly, Vol. LXIX, pp. 323-339.

Engelmann, Joachim. 1976. *German Railroad Guns in Action* Squadron/Signal Publications, Carrollton, Texas

Essame, Hubert. 1969. *The Battle for Germany* Bonanza Books, New York.

Farago, Ladislas. 1981. *The Last Days of Patton.* McGraw Hill, New York.

Fischer, Fritz. 1986. (transl. by Roger Fletcher) *From Kaiserreich to Third Reich — Elements of continuity in German history 1871-1945.* Allen & Unwin, London, Boston, and Sydney.

Fishman, Sarah. 1991. *We will wait — Wives of French Prisoners of war 1940-1945* Yale University Press.

Florentin. Eddy. 1967. (Transl. from French by Mervyn Savill) *The Battle of the Falaise Gap* Hawthorn Books, Inc. New York.

Flynn, George Q. 1993. *The Draft, 1940-1973* Univ. Press of Kansas

Footitt, Hilary and John Simmonds. 1988. *France 1943-1945* Holmes and Meier.

Friedman, Leonard, DDS. (former battalion surgeon/dentist)corresp.1993/1994

Fuks, Ladislav. 1968. *Mr. Theodore Mundstock* (Prague 1942). Orion Press, New York.

GAO (General Accounting Office). 1977. GAO Report to Congress B-146890. The Army's test of one-station training adequacy and value. Dept. of the Army, Feb. 9, 1977.

Garlinski, Jozef. 1985. *Poland in the Second World War* Hippocrene Books, New York.

Geiger, 1993. Stadt Numberg, Naue Postleitzahl 90317, Verwaltungsamt Ost -Postfach 8500, Numberg, Correspondence October 5, 1993. Burgermeister of Fischbach, now a part of Nuremburg.

Gellately, Robert. 1990. *The Gestapo and German Society — Enforcing Racial Policy 1933-1945* Clarendon Press — Oxford.

German Military Studies. 1947. The German Ardennes Offensive The War Room, Inc., LaFox, Illinois

Gillingham, John. 1985. *Industry and Politics in the Third Reich Ruhr coal, Hitler, and Europe* Columbia University Press, New York

Glesinger, Egon. 1942. The impact of war on forest industries. Journal of Forestry 40:6-11.

Gooch, John and Amos Perlmutter. 1982. *Military Deception and Strategic Surprise* (German duplicity after WWI) F. Cass, London.

Goralski, Robert. 1981. *World War II Almanac 1931-1945*

Goutard, A. (Colonel). 1959. *The Battle of France, 1940* Washburn, New York.

Grohne, William (former officer in "A" Btry) Corresp/Disc. 1993/1994 San Francisco, California.

Harmon, E. N. and Milton MacKaye and William Ross macKaye. 1970. *Combat Commander — Autobiography of a Soldier* (Commanding General of the 2nd Armored Division in Europe, and the 1st Armored Division in North Africa). Prentice Hall. Englewood Cliffs, N.J.

Harmon, E. N. 1962-1965. I was assigned as Deputy PMS at Norwich University, Vermont, in 1962. General Harmon was President of Norwich University during the years I was there. We had several discussions on the campaigns.

Harvey, Arthur (former battalion intelligence officer) Corresp. 1994.

Hawes, Stephen. 1975. *Resistance in Europe 1939-1945* Allen Lane (Penguin Books) London.

Hayward, N. F. and D. S. Morris. 1988. *The first Nazi Town — Coburg* St. Martin's Press, New York.

Herneisen, Mrs. Harold. (Arlene) Widow of "B" Battery soldier. 149 Doe Run Rd., Mannheim, PA 17545-9322. (717) 665 4049. Telecon 9/1/93. Herzog, Arthur. 1965. *The War-Peace Establishment* Harper & Row.

Herzog, Wilheim. 1947. *From Dreyfus to Petain — the struggle of a Republic*. Creative Age Press, New York.

Heske, Franz. 1938. *German forestry*. Yale Univ. Press. (p. 144).

Hildebrand, K. 1984. *The Third Reich* (The German buildup for war) Allen and Unwin, London.

Hoffman, Zane (former corporal in "C" Btry) Corresp. & Disc. 1993/1994.

Holbo, Paul Sothe. 1967. *Isolationism and Interventionism 1932-1941* Rand Mcnally, Chicago.

Hoy, George W. 1993. 21 Woodland Drive, Lock Haven, PA 17745 Tel 717 748 6194 Telecon 1993 series.

Infantry Journal Press. 1947. *Conquer — The Story of Ninth Army 1944-1945*. Infantry Journal Press. Washington.

Inman, Verne T., Ralston, Henry J., and Todd, Frank. 1981. *Human Walking* Williams & Wilkins, Baltimore/London.

Instone, Gordon. 1954. *Freedom from the Spur* (Evasion and escape in France, 1940-1941.) Burke, London.

de Jong, Louis. 1990. *The Netherlands and Nazi Germany* Harvard University Press

Journal of Forestry 36:495-503. May 1938. Utilization of wood under Germany's four-year plan.

Kam, Ephraim. 1988. *Surprise Attack — the victim's perspective* (Pearl Harbor and other intelligence failures). Harvard University Press, Cambridge, Mass.

Katti, David. 1989. *A child's war — World War II through the eyes of Children*. Bzztoh, The Hague, Netherlands.

Kardoff, Ursula von. 1966. *Diary of a Nightmare — Berlin 1942-1945* The John Day Company.

Kaufman, Clifford (former officer in "C" Btry) Corresp. & Disc. 1993/1994. Louisville, Kentucky.

Keegan, John. 1982. *Six Armies in Normandy: from D-Day to the liberation of Paris, June 6th — August 25th 1944*. Viking Press, New York.

Kempowski, Walter. 1981. *Days of Greatness* (Germany in the early 1900's) Knopf, New York.

Kennett, Lee B. 1987. *G.I. The American soldier in World War II*. Charles Scribner's Sons, New York.

Kimche, Jon. 1968. *The Unfought Battle* (Why the French and English didn't take the offensive in 1939) Stein and Day, New York.

Knappe, Siegfried and Ted Brusaw. 1992. *Soldat — Reflections of a German Soldier 1936-1949*. Orion Books, New York.

Koch, H. W. 1989. *In the Name of the Volk — Political Justice in Hitler's Germany* St. Martin's Press, New York

Kolakowski, Edmund. 1994 correspondence. Vandergrfit, PA

Kramish, Arnold. 1986. *The Griffin*. Houghton Mifflin Co., Boston.

Kuehn, Heinz R. 1988. *Mixed Blessings — An Almost Ordinary Life in Hitler's Germany* (A half-Jew not imprisoned). The University of Georgia Press, Athens and London.

Lacy, James J. 1994. Personal correspondence. Maugansville, MD.

Laqueur, Walter. 1980. *The Terrible Secret: Suppression of the Truth About Hitler's Final Solution* Little Brown, Boston.

Lehmann, Rudolf and Ralf Tiemann. 1993. *The Leibstandarte IV/1* Fedorowicz Publ. Company. Winnipeg.

Ling, Otto. 1993. 186 Crary Ave., Binghampton, NY 13905. Personal correspondence.

Linton, Derek S. 1991. *Who has the youth has the future — The campaign to save young workers in Imperial Germany*. Cambridge Univ. Press.

Lipstadt, Deborah E. 1986. *Beyond Belief — The American Press and the Coming of the Holocaust 1933-1945* The Free Press, New York/London.

Litoff, Judy Barret et al. 1990. *Miss You — World War II letters* University of Georgia Press.

Lottman, Herbert R. 1985. *Petain, Hero or Traitor: the untold story* William Morrow Company, Inc., New York

Lottman, Herbert R. 1986. *The Purge* (DeGaulle's purge of Vichyites) Morrow, New York.

Lull, Phillip (former battalion survey officer) Corresp/Disc 1994. Phoenixville, Pennsylvania.

Lyons, Gene M. and John W. Masland. 1959. *Education and Military Leadership* Princeton University Press.

Maas, Walter B. 1970. *The Netherlands at War: 1940-1945* Abelard-Schuman, London.

Macksey, Kenneth. 1987. *Military errors of World War Two* Arms and Armour Press, London.

Malmberg, Richard. 1993. 1283 Weisstown Rd., Boyertown, PA 19512. (215) 367 2452. Personal correspondence & telecons

Mann, Klaus. 1984. *The Turning Point — Thirty Five Years in this Century* The Autobiography of Klaus Mann. Markus Wiener Publishing, Inc.

Manning, Paul P. 1981. *Martin Bormann, Nazi in Exile* (Funding of multinational corporations by the Nazis in the closing days of the war) Stuart, Seacaucus, New York.

Martel, Gordon (Ed.) 1986 *The Origins of the Second World War Reconsidered — The A.J.P. Taylor debate after twenty-five years* Allen & Unwin, Boston, London, & Sydney.

Mason, David. 1969. *Breakout, Drive to the Seine* Ballantine Books Inc. New York.

McDonald, Charles B. 1969. *The Mighty Endeavor — American Armed Forces in the European Theatre in World War II* Oxford Univ. Press, N.Y.

McDonald, Hugh. 1945. *Unit History — 283rd Field Artillery Battalion* Printed in France in 1945. No information on printer etc.

MacDonogh, Giles. 1992. *A good German — Adam von TRott su Solz* The Overlook Press, Woodstock, New York.

Markle, Ray (former driver in "C" Btry) Corrsp. 1994.

McKee, Alexander. 1971. *The Race for the Rhine Bridges* Stein & Day, New York.

Medina, Harold (Judge). 1975. *Anatomy of Freedom* (The story of the German spies who landed on Long Island, and their trial.) Holt, New York.

Merton, Robert K. and Paul Lazarsfeld. 1950. *Continuities in social research — "The American Soldier"* Free Press, Glencoe, Ill.

Meuwissen, M. 1993. Series — Personal correspondence from Heemkunde Vereniging Nieuwstadt, Beijerstr. 4, 6118 CS Nieuwstadt, Netherlands.

Middleton, Drew. 1983. *Cross Roads of Modern Warfare* Doubleday. Garden City, New York.

Mierzejewski, Alfred C. 1988. *The Collapse of the German War Economy 1944-1945* The University of North Carolina Press.

Miller, Douglas (U.S. Commercial Attache in Berlin). 1940. *You can't do business with Hitler.* Little Brown & Co., Boston.

Mills, Geofrey T. and Hugh Rockoff (ed.) 1993. *The Sinews of War - Essays on the Economic History of World War II* Iowa State University Press.

Mitchell, Joseph. 1967. *Discipline and Bayonets* Putnam, New York.

Montgomery, Field Marshall the Viscount. 1950. *An Approach to Sanity* World Publishing Company, Cleveland & New York.

vonMonroy, Albrecht. 1938. Deutschlands Helzwirtschaft. Journal of Forestry 36:821-822.

Muller, Ingo. 1991. *Hitler's Justice — The Courts of the Third Reich* Harvard University Press. Cambridge, Mass.

Murphy, James T. 1993. *Skip Bombing* Praeger, Westport, Conn.

Ney, John. 1970. *The European Surrender; a descriptive study of the American social and economic conquest* Little Brown & Company.

O'Neil, James E. & Robert W. Krauskopf (ed.) 1976. *World War II An account of its documents* Howard Univ. Press, Washington, DC

Pappas, George S. 1971. *U.S. Army Military History Research Collection Unit Histories* (Biblio.) Army War College, Carlisle Barracks, Penna.

Palmer, Robert and Bell I. Wiley. 1947. *The Procurement and Training The United States Army in World War II* Series by Office of Chief of Military History., U.S. Government Printing Office.

Parker, James (former sergeant in Service Battery) Corr. & Disc. 1993/1994 Oreland, Pennsylvania.

Pauley, Bruce F. 1981. *The Forgotten Nazis — A history of Austrian National Socialism* University of North Carolina Press

Peifer, William H. 1959. *Supply by Sky* Historical Branch Office of the Quartermaster General.

Pensock, Robert S. 1993. 638 W. Diamond Ave., Hazleton, PA 18201-4936. (717) 455 1676. Peronal correspondence and Telecons.

Pergrin, David E. 1989. *First across the Rhine — The 291st Engineer Combat Battalion* Atheneum Press, New York.

Perkins, Major General K. 1987. *Weapons and Warfare — Conventional Weapons and their Roles in Battle* Brassey's Defense Publishers.

Perret, Geoffrey. 1991. *There's a War to be Won — The United States Army in World War II* Random House, NY

Phillips, Robert F. 1983. *To Save Bastogne* (110th Infantry in the Ardennes) Stein and Day, New York.

Phillips, Robert L. 1984. *War and Justice* Univ. of Oklahoma Press.

Piekalkiewicz, Janusz. 1980. *The Cavalry of World War II* Stein and Day, NY

Powers, Thomas. 1993. *Heisenberg's War — The Secret History of the German Bomb* Alfred A. Knopf, New York.

Ratchford, B. U. and Wm. D. Ross. 1947. *Berlin Reparations Assignment Round One of the German Peace Settlement* The University of North Carolina Press, Chapel Hill.

Read, Anthony and David Fisher. 1988. *The deadly embrace — Hitler, Stalin, and the Nazi-Soviet Pact 1939-1941* WW Norton Co.

Read, Roy S. 1993. 310 Tweedwood Dr., New Haven, IN 46774 (219) 749 8686 Personal correspondence and Telecons.

Rearigh, John E. (former man in "C" Btry). Corresp. & Disc. 1993/1994

Reber, Levan C. 1993. 109 N. Cambridge Ave., Ventnor, NJ 8406 (609) 487 8121 Personal correspondence and Telecons.

Rinderle, Walter and Bernard Norling. 1993. *The Nazi Impact on a German Village* The University Press of Kentucky.

Robbins, Roy M. 1962. *Our landed heritage.* Univ. of Nebraska Press.

Royle, Trevor. 1989. *A Dictionary of military quotations* Simon Schuster.

Sagan, Francoise. 1987. *A reluctant hero* (A love story in the French Resistance). E. P. Dutton, New York.

Sapp, John D. (former battalion operations officer) Corresp. 1993/1994. Leavenworth, Kansas.

Sawicki, James A. *Field Artillery Battalions of the U.S. Army, Vol. 1* (pp. 486-7) U.S. Army Military History Institute. Wyvern Publications, PO Box 188, Dumfries, VA 22026

Schmitz, David F. 1988. *The United States and Facist Italy, 1922-1940* University of North Carolina Press

Schollgen, Gregor. 1991. *A Conservative Against Hitler — Ulrich von Hassell: Diplomat in Imperial Germany, the Weimar Republic and Third Reich, 1881-1944.* (Transl. by Louise Willmot) St. Martin's Press, New York.

Schramm, Percy Ernst, Ph.D. 1946. The Preparations for the German Offensive in the Ardennes (Sep — 16 Dec 44). German Military Studies. The War Room, Inc., LaFox, Illinois. Schramm was the officer in charge of keeping the war diary of the Wehrmacht Operations Staff.

Seaton, Albert. 1981. *The Fall of Fortress Europe 1943-1945.* Holmes & Meier Publ., NY

Shandruk, Pavlo. 1959. *The Arms of Valor* A Spellar, New York.

Sheppard, Morris (Chairman of the Committee on Military Affairs of the United States Senate in 1939) *The Army of the United States* U.S. Government Printing Office. 1940.

Shirer, William. 1960. *The Rise and Fall of the Third Reich* Simon and Schuster, New York.

Shulman, Abraham. 1982. *The Case of the Hotel Polski: an account of one of the most enigmatic episodes of WWII* By Holocaust Library, New York.

Skinner, Michael. 1989. *USAREUR — The United States Army in Europe* Presidio Press, Novato, California.

Slonaker, John. 1972. *The Volunteer Army* (also Universal Military Training) Special Bibliographic Series, Number 5 - A Military History Research Collection Bibliography — Carlisle Barracks, PA.

Stanton, Shelby L. 1984. *World War II Order of Battle* Presidio Press, Novato, CA.

Steenberg, Sven. 1970. *Vlasov—The Role of General Andrey Vlasov in Opposition to Stalin in WWII* Alfred A. Knopf, New York

Stein, George H. 1966. *The Waffen SS—Hitler's Elite Guard 1939-1945* Cornell University Press.

Steinhoff, Johannes (Transl. by J. A. Underwood). 1985. *The Final Hours—A German Jet pilot Plots against Goering* The Nautical and Aviation Publishing Company of America, Baltimore.

Stern, Herbert Jay. 1984. *Judgement in Berlin* Universe Books.

Stithm, Dale D. 1993. 2400 Frontier Street, Longmont, CO 80501 personal correspondence Nov. 12, 1993.

Stoler, Mark A. 1989. *George C. Marshall, Statesman of the American Century* Twayne Publishers.

Stone, I. F. 1988. *The War Years 1939—1945. A non-conformist history* Little Brown and Company, Boston.

Stouffer, Samuel A. et. al. 1950. *The American Soldier* Princeton University Press. Free Press, Glencoe, Illinois.

Strik-Zijlstra, C. E. M. 1993. Personal Correspondence, Sectie Militaire Geschiedenis, Kononklijke landmacht.

Sulzberger, C. L. 1989. *Paradise Regained—Memoir of a Rebel* Praeger, New York.

Sweetman, John. 1982. *Operation Chastise—The WWII busting of the Ruhr River Dams* Janes Publishing Company, London.

Taylor, George Rogers. 1971. *The Turner Thesis.* D.C. Heath & Co.

Taylor, Robert L. and William E. Rosenbach (Ed.). 1984. *Military Leadership* Westview Press, Boulder, Colorado.

Thompson, Laurence. 1968. *The Greatest Treason; The untold story of Munich* William Morrow Company, New York.

Thompson, R. W. 1968. *D-Day—Spearhead of Invasion* Ballantine Books Inc., New York.

Travers, Timothy and Christian Archer. 1982. *Men at War* (Politics, Technology and Innovation in the Twentieth Century) Precedent, Chicago.

Treitschke, H. G. von 1969. *Origins of Prussianism* H. Fertig, NY.

Truscott, Lucian K. Jr., General. 1989. *The Twighlight of the U.S. Cavalry 1917-1942* University Press of Kansas.

Turner, Frederick Jackson. 1920. *The frontier in American history.* Henry Holt & Co.

Venzon, Anne Cipriano. 1992 *General Smedley Darlington Butler -The Letters of a Leatherneck, 1898-1931.* Praeger, New York

Virheyen, Dirk and Christian Soe (ed). 1993. *The Germans and their Neighbors* Westview Press, Boulder, San Francisco

Wagenmann, William. 1993. 1216 Clayton Ave., Oakford, PA 19053-3607 (215) 357 5291 Personal telecons.

Wardach, Stanley (former man in "C" Btry). Corresp/Disc 1993.

Watts, Meredith W. 1989. *Contemporary German youth and their elders* Greenwood Press, New York.

Weber, Wilhelm and Eduard Weber. 1894. (Reprint 1992 — Springer-Verlag) *Mechanics of the Human Walking Apparatus*, Verlag von Julius Springer

Weighton, Charles and Gunter Peis. 1958. *Hitler's Spies and Saboteurs* (Based on war diary of- General Lahousen) Henry Holt & Co., New York

Weigley, Russell F. 1981. *Eisenhower's Lieutenants — The Campaign of France and Germany 1944-1945* Indiana University Press.

Weinberg, Gerhard L. 1981. *World in Balance* (Behind the scenes of World War II. Hitler's 1920's plan for war, his image of the U.S., German colonial plans, and Munich). University Press of New England.

Weizsacker, Richard von. (President of West Germany). 1987. Translated by Karin von Abrams. *A Voice from Germany — Speeches by Richard von Weizsacker* Weidenfeld & Nicholson, New York.

Wheal, Elizabeth-Anne et al. 1989. *Encyclopedia of the Second World War* Castle Books,

Whealey, Robert H. 1989. *Hitler and Spain — The Nazi role in the Spanish Civil War 1936 -1939* University Press of Kentucky.

Whitaker, W. Dennis and Shelagh Whitaker. 1989. *Rhineland — The Battle to end the war.* St. Martin's Press, NY

Whitcomb, John and Claire Whitcomb. 1987. *Oh Say Can You See — Unexpected Anecdotes about American history* William Morrow and Company.

Whiting, Charles. 1990. *The Other Battle of the Bulge — Operation Northwind* Scarborough House Publishers, Chelsea, MI 48118

Willeford, Charles. 1986. *Something about a soldier— the 1930's Army* Random House, New York

Willis, Donald J. 1988. *The Incredible Year* War diary with the 67th Field Artillery Battalion. Iowa State Univ. Press, Ames.

Willis, James F. 1982. *Prologue to Nurnberg* (The politics and diplomacy of punishing war criminals os World War I, including the Kaiser). Greenwood Press. Westport, New York.

Wilmot, Chester. 1952. *The Struggle for Europe* Collins, London.

Wilmot, Chester. 1989. *The Great Crusade* The Free Press, NY

Wilson, John B. 1987. *Armie, Corps, Divisions, and Separate Brigades* Army Lineage Series, Center for Military History, Washington, DC

Witt, Fred. 1945. Informal notes written during the war. Partial diary.

Woirol, Gregory. 1992. *In the Floating Army — F.C. Mills on Itinerant Life in California in 1914* University of Illinois Press.

Wolf, Herman (fomer officer in "C" and Service Battery) Disc. 1994.

von Zangen, Gustav. 1947. Fifteenth Army Defense Battles at the Roer and Rhine (22 November 1944—9 March 1945). German Military Studies. (After-action report) The War Room Inc., LaFox, Illinois.

de Zayas, Alfred Maurice. 1993. *German Expellees: Victims in War and Peace* St. Martin's Press, NY

Zurndorfer, Hannele. 1983. *The Ninth of November 1938* (The story of a Jewish girl's escape to England along with evidence that the vast majority of Germans were supportive of the Nazis.) Quartet Books, London.